PORTRAITS *of the* AFRICAN WILD

Published by
Chartwell Books, Inc.
A Division of Books Sales, Inc.
110 Enterprise Avenue
Secaucus, New Jersey 07094

© Photography Gerald Cubitt
© Text Mike Nicol
Book Editing: Allan Duggan
Book Design: Joan Sutton
Reproduction and Typesetting: Hirt & Carter (Pty) Ltd. Cape Town

First published in the United States of America 1986 by Chartwell Books, Inc.

ISBN 1-55521-043-0

Printed by Cayfosa, Barcelona, Spain
Dep. Leg. B - 2575 - 1986

PORTRAITS *of the* AFRICAN WILD

Gerald Cubitt

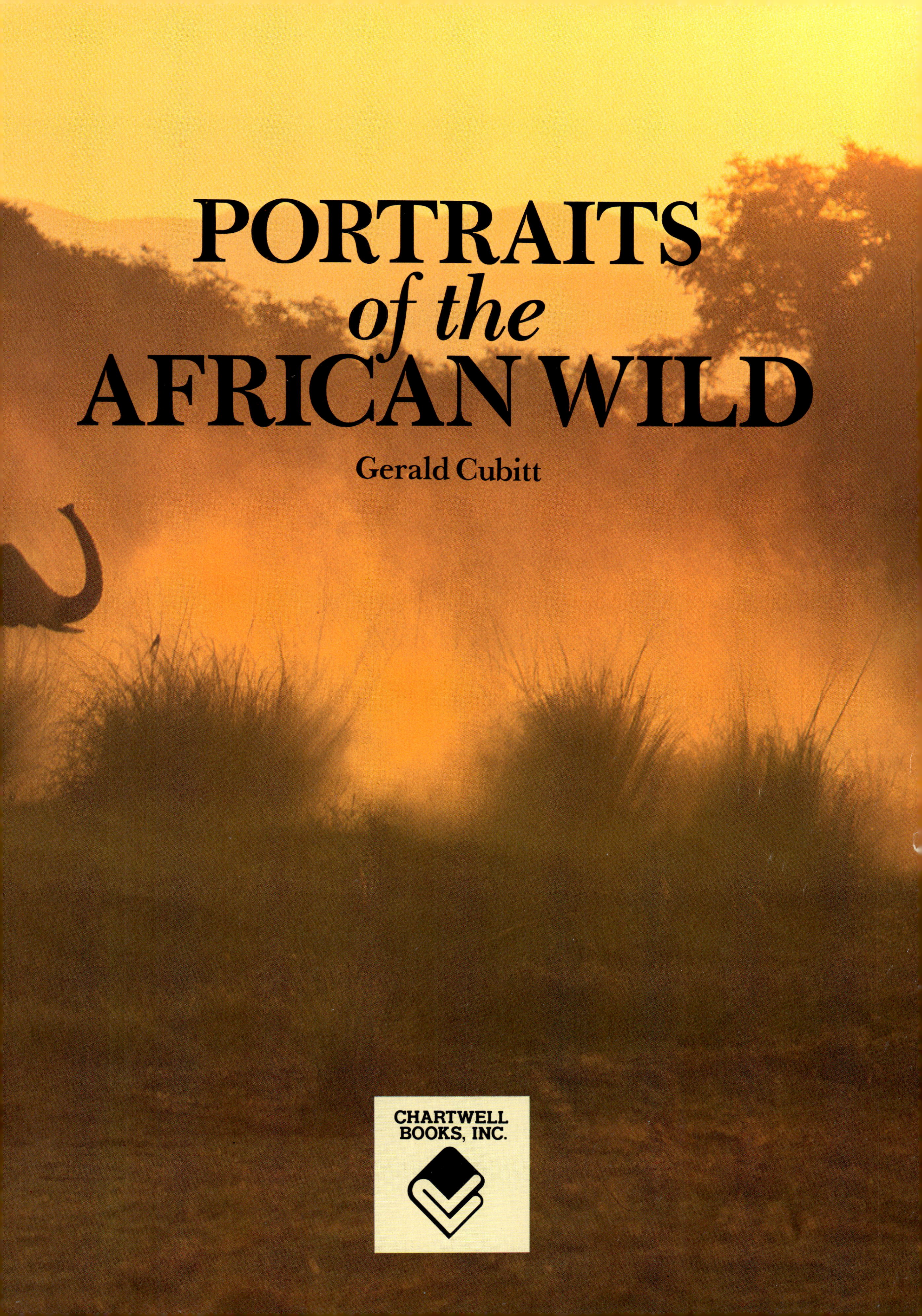

CHARTWELL
BOOKS, INC.

WWF WORLD

With each passing year, our magnificent natural heritage appears to face new threats. The ever-widening ripples of humanity are lapping at the edges of wild places which have been virtually untouched for aeons, and in many places have overwhelmed, tamed and utterly destroyed ecosystems and wildlife.

Today, only a dedicated few stand between the relentless demands of 'progress' and the beleaguered sanctuaries – but they are growing in number. Their collective conscience has prompted far-reaching programmes aimed at the salvation of this natural bounty.

Among those who care are people like Gerald Cubitt, whose photographs clearly reflect his love and understanding of wildlife. In PORTRAITS OF THE WILD, he captures the essential Africa – raw, unspoilt and quite beautiful.

The World Wildlife Fund acknowledges and admires Gerald Cubitt's skill in capturing animals on film as they are in life, portraying large and small mammals and habitats in their true splendour.

It is fervently hoped that the photographs in this book will win new converts to the cause of conservation and, perhaps, provide new delights for the already converted.

Anton Rupert

Anton Rupert
Vice President: World Wildlife Fund

The Symmetry of Survival

Five hundred years ago, the southern tracts of the African continent teemed with a seemingly inexhaustible number and variety of wild animals. Great herds of buffalo moved slowly over savanna plains; vultures patrolled the skies, waiting to spiral down on abandoned kills. In the long grass, tawny lions stretched out, replete. All living things were part of a grand and beautiful design, living, reproducing and dying in cycles that had changed little in thousands of years. Man was part of it. He was just another predator, albeit armed with crude stone weapons, and just as vulnerable as any other animal to disease, drought, or a bigger predator. In harmony, if not peacefully, he lived among the greatest variety of mammals on any land mass in the world.

Nature had produced differed habitats which supported different species. Herbivores existed in mutual co-operation: one species grazed long grass, enabling another to feed off the short stems. Yet ecosystems were never destroyed. Nature's 'programme' was perfectly balanced with climatic, parasitic or predatory checks and controls. Only the fittest survived, ensuring that only the best genes were passed from generation to generation. If there was a flaw in nature's grand design, it was Man. As his brain evolved, he became able to alter his natural environment in a way no other animal could. Long before the European visitors made their inroads into the unspoilt wilderness, the land had felt the first impact of humanity's ever-growing demands. Lands had been put to the plough, cattle to pasturage.

However, it was the coming of the men in tall ships that did most to change the face of Africa. They brought the drive to develop, to transform the veld into farmland and cities, to dam rivers and erect fences. Nothing, it seemed, could slow the relentless tide of people that swept deep into the hinterland. In the face of this 'progress', wild animals were clearly an obstacle. They were hunted – some to extinction, some to the very edge. Worst of all, their habitats were destroyed. As the frontiers were pushed outwards, the animals retreated. Fortunately, Man came to his senses in time to create sanctuaries for their dwindling numbers, recognising a heritage he had come close to losing.

Even now, with conservation and environmental consciousness at a peak, the problems are far from over. Nature cannot run a game reserve. In time measured by a 'biological clock', the natural checks and balances would ensure its survival. However, as *we* measure time, it would need only a few short years to transform a lush park into an over-grazed wasteland in which only small creatures would be likely to survive. So the veld and the animals it supports must be managed, using the best scientific means available. Management techniques include burning the veld, creating artificial water holes and culling herbivore populations when they begin to make too serious an impact on the vegetation. Southern Africa is unique: it is Africa's treasure house. Nowhere else on the continent can such a variety of wildlife be seen against such diverse landscapes. There is the barren western desert, where a herd of elephant lives in arid, rocky terrain. The unique Okavango delta, the only place in the world where a river ends inland, is perhaps the last truly wild realm. There are South Africa's famous national parks, protecting such vulnerable species as bontebok and mountain zebra, and there are the vast open stretches of bush and countless animals that make Zimbabwe an animal-lover's dream. As a wildlife package, the region has immense potential for generating foreign exchange. This might be a telling factor in the world of tomorrow, in which economic considerations are likely to carry more weight than moral or aesthetic principles.

There is another, equally valid reason for conservation. In some distant decade, Man might need to draw again from the natural habitats, just as he once drew on them for crops and cattle. More practically, wild animals have long been of use to medicine, and some plants produce chemicals which are difficult to synthesise. It is widely believed that the maintenance of this planet's genetic diversity is a matter of human survival. Weaken the broad genetic base, and we could drastically alter a symmetry we are only beginning to understand.

Mike Nicol

Previous three spreads:
A huge herd of buffalo roams across the wilderness. Wild, vast, criss-crossed with game trails, the unique Okavango Delta retains the character of an Africa that is steadily disappearing under the growing onslaught of mankind.

A lion with his mate in the bushveld of Timbavati Game Reserve.

An impala herd grazes peacefully beside the shimmering blue waters of Zimbabwe's Kariba Dam.

Right: A chacma baboon in the Kruger National Park.

Following pages:
Ostriches file past a white rhino and its calf in the lush grasslands beside Loskop Dam.

As dusk approaches, elephants and buffaloes gather at the Caterpillar Pan in the Hwange National Park.

ANGOLA
Kunene mouth
Kunene river
Okavango river
CAPRIVI GAME RESERVE
Caprivi Strip
Linyanti
MOREMI WILDLIFE RESERVE
CHOBE NATIONAL PARK
Kaokoveld
ETOSHA NATIONAL PARK
NAMUTONI
TSUMEB
Okavango Delta
MAUN
SKELETON COAST PARK
WATERBERG PLATEAU PARK
TORRA BAY
NAMIBIA
CENTRAL KALAHARI GAME RESERVE
CAPE CROSS
WINDHOEK
SWAKOPMUND
NAMIB-NAUKLUFT PARK
BOTSWANA
MABUASEHUBA GAME RESERVE
GEMSBOK NATIONAL PARK
KALAHARI GEMSBOK NATIONAL PARK
Namib Desert
KEETMANSHOOP
LUDERITZ
ROB FERREIRA GAME RESERVE
UPINGTON
Orange river
AUGRABIES NATIONAL PARK
ALEXANDER BAY
SPRINGBOK
SOUTH AFRICA
ROLFONTEIN NATURE RESERVE
AKKERENDAM NATURE RESERVE
VICTORIA WEST
CLANWILLIAM
KAROO NATIONAL PARK
TSITSIKAMMA COASTAL NATIONAL PARK
BONTEBOK NATIONAL PARK
DE HOOP NATURE RESERVE
CAPE TOWN
CAPE OF GOOD HOPE NATURE RESERVE

Zambezi river
MALAWI
BLANTYRE
TETE
MANA POOLS NATIONAL PARK
Lake Kariba
Bumi Hills
MATUSADONA NATIONAL PARK
A FALLS
AL PARK
CHIZARIRA NATIONAL PARK
HARARE
UMA PAN NATIONAL PARK
NYANGA NATIONAL PARK
ZIMBABWE
WANGE
ATIONAL
PARK
GORONGOZA NATIONAL PARK
MUTARE
Rio Zambeze
GWELO
BULAWAYO
FORT VICTORIA
LAKE KYLE
CHIMANIMANI NATIONAL PARK
BEIRA
GADI
ESERVE
MATOPOS NATIONAL PARK
GONAREZHOU NATIONAL PARK
Limpopo river
VENDA
LANGJAN NATURE RESERVE
NWANEDI NATIONAL PARK
KRUGER NATIONAL PARK
MOZAMBIQUE
PIETERSBURG
TIMBAVATI NATURE RESERVE
LONDOLOZI GAME RESERVE
MALA MALA GAME RESERVE
PILANESBERG NATIONAL PARK
LOSKOP DAM NATURE RESERVE
SKUKUZA
PHUTHATSWANA
PRETORIA
JOHANNESBURG
MBABANE
MAPUTO
MLILWANE WILDLIFE SANCTUARY
SWAZILAND
KOSI BAY
NDUMU GAME RESERVE
MHOF DAM
RE RESERVE
NEWCASTLE
LAKE ST LUCIA
WILLEM PRETORIUS GAME RESERVE
MKUZI GAME RESERVE
HLUHLUWE GAME RESERVE
UMFOLOZI GAME RESERVE
GOLDEN GATE NATIONAL PARK
ROYAL NATAL NATIONAL PARK
MASERU
LESOTHO
SEHLABATHEBE NATIONAL PARK
DURBAN
e river
TRANSKEI
UMTATA
CISKEI
IES VOSLOO
U RESERVE
EAST LONDON
THOMAS BAINES NATURE RESERVE
T ELIZABETH

For Janet,
who shares my love for the wilds of Africa.

Contents

Big Game

About 60 million years ago, a small, pig-like amphibious creature scrabbled for grubs in the prehistoric undergrowth. A primitive animal, yet one for which evolution had a crucial role. During the next 20 million years, this species evolved into the massive mastodon, with its sweeping, 2m-long tusks.

Evolution was still unsatisfied. The mastodon made way for the stegodan. The fossil record shows this creature roamed over the entire world except for Australasia and South America. It was ancestor to three groups: the extinct mammoth, the Asiatic elephant and the African elephant.

Today, the African elephant is the world's largest living land mammal. Reaching a height of three to four metres, with a wrinkled, slightly hairy skin which seems to fit badly, this giant has become the symbol of Africa's dwindling natural heritage.

The once great herds have fallen to the relentless guns of hunters and poachers. And as Africa's human population expands, so the wild domain shrinks. Paradoxically, more of Africa's three million elephants wander free than are to be found inside the protective fences of parks and reserves. Not that fences guarantee long life: the lure of big earnings from ivory attracts all too many poachers. Even in the Kruger National Park, considered one of Africa's foremost wildlife sanctuaries, over 120 elephants have fallen to poachers, many of them using automatic weapons.

These massacres are often a symptom of rural poverty, yet it is said that starving villagers could hardly be blamed for poaching when one considers that the price of a tusk could feed the entire village for a year. Inevitably, it is the elephants carrying the biggest tusks that appear in their sights. Many of the continent's biggest tuskers are to be found in the Kruger National Park.

Until recently, the Park boasted an undisputed giant. Named Mafunyane, 'the one who is easily annoyed', he wandered freely through the Park's remote northern regions, and was seldom seen. On his death – through natural causes – in late 1983, his tusks were found to weigh 75 kg and 85 kg; they were three metres in length. There are other big ivory carriers in the wild north. Like Mafunyane, they are usually accompanied by their 'askaris' – younger bulls that, according to legend, protect the old giants. Many an elephant hunter seeking an impressive trophy has had to contend with an askari and been forced to abandon his target.

Elephants are nature's bulldozers. Their passage across the veld is marked by broken branches devoid of leaves and tree trunks stripped of bark. Whole trees are pushed over to expose juicy, tender roots. The elephants' impact upon their environment is awesome; if confined to too small a habitat, it can be catastrophic.

In the days when game wondered freely, nature would repair the damage – given time. But even in the largest game reserve, elephants cannot be left to reign unchecked: they would create such havoc that their habitat – and that of countless other animals – would soon be ruined.

Consequently, these areas must be managed. The most dramatic management technique, and the one that has prompted considerable public controversy, is culling. To many people, culling is merely a licence to massacre. To those who must perform this unpleasant but necessary task, it is a highly scientific and efficient operation.

Family groups of six to ten animals are selected and darted with a powerful drug by marksmen firing from helicopters. Once the drugged animals fall, they are dispatched with a brain shot. On average, these operations take no longer than 20 minutes. The carcasses are butchered in the veld, and in most cases the money earned from the sale of the meat and other products is ploughed back into conservation.

Such drastic measures have to be taken with the big herbivores, such as elephant, buffalo and hippo, because of their impact on the habitat. An adult elephant, for example, eats between 100 kg and 200 kg of vegetation daily, and might drink 90-120 litres of water – enough to fill four bathtubs.

In arid areas, and during droughts, elephants help provide water for other species by digging in dry river beds with their toes, tusks and trunks to reach underground water. They then draw the slowly oozing liquid into their trunks and squirt it into their mouths. Although these huge creatures drink once a day when water is readily available, they can go for up to three days without water.

Elephants are rough feeders. Their diet of marula fruits, mopani leaves and bark, acacia seed pods, palm fruits and grasses places heavy demands on their teeth. Consequently, the molars are replaced six times in the animal's life span of 50-70 years. The molars, known as M1-M6, develop on each side of the upper and lower jaws throughout the elephant's life. They are useful in determining the creature's age: as each tooth wears, fragments break off and the tooth behind it moves slowly to the front, as if on a conveyor belt. When the last set has worn down, the elephant faces the unpleasant prospect of death by starvation.

An elephant's first two sets of molars are usually gone by age six, the third set lasts until the elephant is 15 years old, the fourth set until age 28 and the fifth until about age 47, when the sixth and final set appears.

Elephant society is matriarchal. This means there is a female leader in a family group, or a number of female leaders when groups form a herd. Young males within these groups leave when they reach puberty, at about 10 years of age.

The family group consists of an adult female and her calves. She might be joined by a number of closely related females with their calves, and sometimes these family groups may form herds of between ten and 50 members. It is only when a female is in oestrus that bulls may join the family herds; once the mating season is over, they rejoin their 'bachelor' herds or go back to their solitary ways. Certainly there are no strong bonds between bull elephants.

The calves in a family group may range in age from one to 16 years. When there is plentiful vegetation and water, cows may give birth to a single calf every four years after a gestation period of 22 months.

Births are usually attended by a number of females and their calves, which act as guards to the vulnerable cow. As the calf is expelled, the umbilical cord breaks. The mother gently peels off the foetal membrane with her trunk, tusks and forefoot. At birth, a calf weighs anything from 90-120 kg, and may be up to 1m tall. Within half an hour, it will have struggled to its feet to suckle. Because the young are dependent on their mother for about five years, a cow may suckle more than one calf at a time.

Elephants in a family group communicate with one another in a number of ways. They may touch by rubbing bodies and heads together or by stroking each other with their trunks, sometimes entwining them. When on the move, or eating, they keep contact by a loud rumbling which is often mistaken for the workings of the animal's digestive system. In reality, the rumbling emanates from the throat. When an elephant detects danger, it immediately stops rumbling to warn others. They then stand still and test the air with their trunks.

Like all animals, elephants have an essential role in the balance that maintains an ecosystem. Their dung helps seeds to germinate, and returns nutrients to the soil – a valuable link in the delicate balance of nature.

To look at the future of elephants in Africa is also to look at the future of wildlife in general. Tragically, elephants are incompatible with Man's agricultural demands, and it is merely a matter of time before they will all be restricted to game reserves.

With human numbers on the continent growing at a staggering rate, it is conceivable that without a drastic trimming of the birth rate in the next decade, there may soon be room for neither elephants nor conservation areas. It is becoming increasingly obvious that more effective planning of land-use must be initiated by all African governments to ensure total utilisation and optimum benefit from all wildlife and ecosystems.

In the end, economics will decide the fate of Africa's wildlife. Food and revenue demand high priorities: if wildlife can supply both protein and foreign exchange on a sustained yield basis, then conservation will attain wide acclaim. In short, wildlife must pay for its continued existence. Elephants are major tourist attractions, but they could also be a major food source. Given protection and careful management, their numbers would increase to the point where culling is necessary. Under controlled conditions, hunters could be charged to shoot the excess animals and the meat given to local people.

At the beginning of the 19th century, Africa teemed with large herds of animals. Then came the white man with his gun. Animals were shot for hides, trophies, meat. Among the favourite targets were the black rhinoceros and white rhinoceros. Their horns were highly prized in the East, where they were reduced to powder and consumed in the mistaken belief that they were aphrodisiacs. This belief still persists, and is one of the reasons why great numbers of rhino have been killed in the last three decades.

The white rhino, the bigger of the two species found in southern Africa, is the second largest land mammal. Males can weigh up to 2½ tons, compared to the average weight of one ton for the black rhino.

Despite their names, rhino are generally coloured grey. 'White' is probably a linguistic corruption by early hunters and explorers of the Afrikaans word 'wyd', meaning wide. This refers to their wide mouths. The origin of the black rhino's name could probably be traced to an early explorer who came across a rhino that had been wallowing in black mud. Rhino frequently roll in mud and dust to rid themselves of ticks and other parasites; so they can become whitish, yellowish or reddish, according to the colour of the soil.

In 1652, Jan van Riebeeck recorded in his diary that rhino were found on the slopes of Table Mountain and on the Cape 'Flats'. These must have been black rhino, as the other species was seen for the first time at the end of the 18th century, when pioneers crossed the Orange River.

Black rhino must have once roamed from the Cape northwards, although they were never at home on the Transvaal highveld, or in the arid Kalahari. By the 19th century, they had been exterminated in most of their southern African range. Black rhino became extinct in the Free State in 1842; the last Cape rhino was shot near Addo in 1853. About a century later, they were also extinct in the Transvaal. In Natal, small populations survived in the Umfolozi, Hluhluwe and Mkuzi game reserves, but these numbered a mere hundred in 1930. The present population was built up from that stock, and the species was reinstated in the Kruger National Park.

Black rhino are thriving in Etosha, but the same cannot be said for other parts of Namibia. Poaching has knocked back numbers in Kaokoland and Damaraland to less than 30 individuals. Without adequate protection, the remaining few will go the same way as those in East Africa, where poaching is rife.

In Zimbabwe, black rhino are confined mostly to the shores of Lake Kariba, the Zambezi valley and the Sabi valley. They have also been reintroduced to Hwange. There are probably no more than 800 of these animals left in that country; they are an endangered species.

The dismal litany of extermination has been repeated among white rhino. This species was first recorded by William Burchell early in the 19th century (at Kuruman, near Kimberley). White rhino were subsequently discovered in Natal, the Free State, western Transvaal and Botswana. Before long, however, they were to be found only between the Black Umfolozi and White Umfolozi rivers in Natal. They had been eliminated by bullet everywhere else.

Their survival in Natal was largely attributable to the tsetse fly. Tsetse flies carry the dreaded *nagana*, the sleeping sickness that kills men and cattle. Because the area between the two Umfolozi rivers was seriously infested, hunters were mostly unwilling to risk their lives. But conservationists felt the need for legal protection: in 1897, the area was proclaimed a reserve by the governor of Natal. The last remaining white rhino were safe.

Safe from hunters, perhaps, but hardly safe from extinction. In 1930, there were a mere 28 white rhino left. But the numbers gradually increased, and by 1965 the Natal Parks Board was beginning to distribute the animals to reserves throughout southern Africa, ensuring that the population was no longer confined to one potentially vulnerable region. Today, the white rhino is no longer considered an endangered species.

One of the main reasons these creatures were diverted from the path to extinction was because of a drug called M99. Using this drug, rhino could be tranquillised, loaded into trucks and carried long distances to other reserves or zoological gardens.

One recipient of white rhino was Kenya's Meru National Park. With rhino horn trading in the mid-1980's at between U.S.$500 and $750 an ounce, poachers have had a powerful incentive to break the law: it is estimated they have wiped out about 90 per cent of central Africa's rhino population. In six years, Kenya lost 10 000 black rhino, and there are now fewer than 1 500 left. During 1975-1976, about 8 300 kg of rhino horn were exported to North Yemen from Kenya.

The simplest way to distinguish between the two species is to look at the shape of their heads. The head of a white rhino is long and broad, with a square mouth which suits its grazing habits. The black rhino is a browser, with a shorter, bullet-shaped head and a prehensile lip. The difference in their eating habits explains why white rhino prefer the grassy plains, while black rhino are found in bush country.

Behavioural differences between the two species are quite marked. White rhino occur in small groups characterised by a territorial bull, subordinate bulls, cows and their calves. The territorial bull marks the home range by spraying urine along the boundaries, or by dung heaps. However, dung heaps also occur elsewhere within his territory. Subordinate bulls spend most of their lives within the dominant bull's home range, provided they do not challenge him. They do not spray urine or patrol the boundaries; nor do they mate with the cows. Cows are free to wander inside the marked territory.

Black rhino are more solitary, and not as territorial. Like the territorial bulls of the other species, the males tend to avoid each other, but bulls may fight over a cow that is ready to mate – a contest which could end in death for one of them. When fighting, they squeal and knock one another with their front horns.

Apart from Man, rhino have no enemies – though their best friends have been known to cause them fatal injuries. Oxpeckers, which have a ferocious appetite for ticks, scour the backs of rhinos, giraffes and other animals. Mostly they do their hosts a service, but on occasion they pick off ticks so furiously that they open small wounds. Relentlessly, these birds keep on and on at the festering sore, until the stricken animal finally dies.

An anecdote from the Kruger National Park recalls the time a ranger came across a rhino leaning its head against a tree trunk. Hours later, when the ranger passed that way again, he noticed that the animal was in the same position. The ranger sat and waited. Eventually, the rhino moved off – revealing a gaping hole in its head. Because it was obviously in pain, and unlikely to recover from the festering wound, the ranger shot the rhino. A post-mortem revealed that the wound had been caused by an oxpecker: the rhino had presumably been leaning against the tree to prevent the bird from further irritating the wound.

While both species of rhino have poor eyesight, their hearing and sense of smell are acute. Yet they differ once again in gestation periods, which is 19 months for white rhino, against 17 months for black. Females of both species reach sexual maturity between the ages of four and five years and give birth every three years. Both species give birth to a single calf. An Interesting difference between the two species: the white rhino calf leads its mother, whereas the black rhino calf is led. Rhinos live for about 40 years.

A fallacy which has often proved the undoing of hunters concerns the rhino's supposed lack of agility. In fact, they can charge at speeds of up to 50 kilometres an hour, and despite their bulk, can turn remarkably quickly.

Wounded animals, or animals that feel threatened, can be highly dangerous. Wounded buffaloes, for example, have been known to double back and 'ambush' their attackers. Many a careless hunter has come to grief on their massive, bossed horns.

The famous hunter-explorer of the last century, Frederick Courteney Selous, personally shot about 200 buffalo. In his book, *A Hunter's Wanderings in Africa*, he described many of his encounters with these animals. Time and again, he tells of the buffalo's dangerous determination. One incident almost resulted in the death of his gun-carrier. It was in the long grass near Chobe that they came across an old bull lying down.

'My friend gave him a bullet as he lay,' recounts Selous, 'upon which he jumped up and stood behind some mopani trees, only exposing his head and hind-quarters on either side of their stems. After eyeing us for a few seconds, he turned and went off at a gallop, but before he had gone many yards Wood (the friend) fired at him with his second gun and knocked him over; he was on his legs again in a moment, and wheeling round came straight towards me at a heavy gallop, his nose stretched straight out and grunting furiously.

'When he was about twenty yards from me, I fired with my large four-bore elephant gun, and struck him fair in the chest.' This staggered but did not stop the buffalo, wrote Selous. 'Swerving slightly, it made straight for the gun-bearer, who promptly threw down the second gun and commenced climbing a tree. The buffalo scraped its right horn up the trunk, catching the gun-bearer a severe blow on the inside of his knee and nearly knocking him out of the tree.' Added Selous: 'The sturdy beast then ran about twenty yards farther, knelt gently down, and stretching forth its nose commenced to bellow, as these animals almost always do when dying; in a few minutes it was lying dead.'

Despite their poor eyesight and hearing, buffalo go virtually unchallenged in the wild. Lions might prey on the old, solitary, sick and young, but otherwise give buffalo a wide berth; they have learned that buffalo defend themselves and their young with awesome vigour. However, crocodiles occasionally attack a calf, as do spotted hyenas.

When threatened, the bulls force the cows and calves into a group while they turn to face the danger. An alarmed herd will gallop off at a thundering pace, flattening everything in a headlong stampede. Unmolested buffalo, on the other hand, are generally as docile as cattle.

Buffalo are gregarious creatures. Herds sometimes number as many as two thousand, although such vast gatherings are rare today. Being dependent on a ready supply of water, they are never far from rivers or water holes, and are most likely to be found in heavy bush country with nearby grassy plains. Although they sometimes browse on shrubs, they live mainly on tufts of long, coarse grass. In this way, they clear an area for other grazers such as zebra and wildebeest, which prefer shorter grass. Buffalo may have poor eyesight and hearing, but their sense of smell is highly developed: they use their damp, sensitive noses to detect danger as well as to distinguish between edible and poisonous plants.

Interactions among wild animals form complex jigsaw puzzles. For example, where there are giraffe there might also be zebra, wildebeest and impala. These animals associate with one another to increase their chances of survival. To them, the giraffe is a living 'lookout post'.

Most adult giraffe are about 5 m tall. They have excellent vision and hearing, enabling them to detect predators a long way off. Their height has also given them a special ecological niche, allowing them to feed on trees which other species cannot reach – a considerable advantage in times of drought.

When browsing, the giraffe's long tongue curls around twigs to reach the leaves. Its long, narrow mouth also aids it in plucking leaves among acacia thorns. Thorns pose no problems for the giraffe, as the roof of its mouth is covered with a horny skin. Its thick and rubbery saliva also helps by coating the spikes with a protective layer. The nostrils can be closed to protect them from twigs and thorns.

Despite the giraffe's characteristically long neck, it has only seven vertebra – just like humans and most other vertebrates. In the giraffe, however, they are greatly extended and jointed to give maximum flexibility. Not surprisingly, the giraffe has one of the largest hearts in the animal kingdom: it weights 12 kilograms. This big heart is essential, as it needs to pump blood about 3 m uphill to the brain. The jugular vein, connecting the heart to the brain, is 2 cm in diameter and lined with special valves to prevent blood from flooding back to the heart and lungs. When the animal stoops to drink, these valves prevent the blood from rushing to the head.

Giraffe might have a docile appearance, but bulls often fight – using their necks and heads. Standing side by side, they pound each other with their heavy knobbed heads. The combat may be alarming, but fights seldom end in death. There is no obvious winner; the loser simply moves away when it has had enough.

While bulls tend to be solitary, cows live in social groups with their young. During the day, the young are herded into a 'nursery' group watched over by one or two adult females, known as 'aunts'. Other females are then free to feed.

Lions are the giraffe's major enemy. But even lions have been known to abandon this tall prey for easier kills; giraffe protect themselves and their young by kicking viciously.

Burchell's zebra is one of the most common animals in the game parks of southern Africa. Usually found in small herds, or mingling with other animals, they are a symbol of the vast savanna plains.

Zebra lead a herd life: the herd is either a family or a bachelor group. A family group is headed by a stallion with six to ten mares, foals and yearlings. Bachelor groups vary from two to ten animals.

Members of a family group all have their place. When the family is on the move, it is led by a dominant mare, followed by lower-ranking mares. Foals follow the mother in order according to age; the youngest first.

As a rule, the lead mare decides where the family will graze and drink, though the stallion – which walks to one side of the group – can change the direction. His role is to protect the family and keep the foals from wandering off.

Unlike many other plains animals, zebra walk or run at the speed of the slowest in the herds. Even when being hunted, they will not abandon a wounded, sick or old member of the group. Occasionally, an old or sick stallion in a family group will leave to live on its own.

Fights may occur within a family group when a colt tries to 'steal' a mare, but these contests rarely have fatal results. The combatants bite at each other's neck, throat, mouth and forehead, sometimes standing on their hind legs to 'box'.

Within 30 minutes of being born, a foal struggles to start walking. Once it can stand, it begins to suckle, and after ten days will be grazing –

though it continues to suckle for its first year. About half of the zebra foals survive to maturity; the rest die from disease, lack of food or predation.

While their main enemies are lions, packs of hyenas and wild dogs, leopards and cheetahs also attack foals. Bringing down a zebra – especially a strong and vigorous male – could have severe consequences for the predator. Like the giraffe, the zebra has a kick so powerful that it has been known to smash a lion's jaw. It can also inflict serious injuries with its formidable teeth.

Burchell's zebra is not the only zebra species found on the sub-continent. In the dry mountainous regions of the Karoo, and in Namibia, are found the Cape mountain zebra and the sub-species Hartmann's zebra. Hartmann's zebra is more horse-like than the Cape mountain zebra, with widely spaced, narrower rump stripes.

There are many differences between the Burchell's zebra and the Cape mountain zebra. The Cape mountain zebra is smaller and more compact, with larger ears, a distinct dewlap and a red-brown muzzle. They also have clear black and white stripes with no shadow marking, and the legs are striped to the hoof. The bellies are white, and there is a characteristic grid-iron pattern on the rump.

As the name implies, the Cape mountain zebra prefers dry, rocky terrain, although it is also a grazer. To cope with the abrasive effect of the ground, their hooves grow continuously to make up for wear and tear.

The Cape mountain zebra is among the rarest mammals on earth. At the turn of the century, this species – like some other indigenous animals – was heading rapidly for extinction. It appeared that the authorities could see no reason to institute protective measures.

By 1930, Cape mountain zebra had been virtually wiped out. Then, in 1937, the National Parks Board proclaimed the Mountain Zebra National Park near Cradock, in the Cape. At the time it contained only six zebra.

Ironically, these animals disappeared within a few years, but fortunately, a neighbouring farmer came to the rescue. In 1950, he donated 11 zebra, a nucleus from which was built a herd of over two hundred. This represents about 94 per cent of the Cape mountain zebra population in South Africa.

No other mouth in the wild Kingdom is quite as cavernous or as 'plush' as that of the corpulent hippo. It opens its mouth to reveal rich ivory teeth cushioned in what appears to be vermilion silk upholstery. The hippo's teeth grow continually to make up for the eroding effects of its rough vegetarian diet.

When a hippo opens its mouth to 'yawn', it is not because the hippo is tired: it is actually threatening an adversary. Why hippo fight is not quite clear, but it is known that they do so more often and more savagely than most other animals. Solitary old bulls are aggressive, and will attack cows and young calves. They even attack river boats and have been known to kill humans.

To the Greeks, hippo were 'river horses', and indeed they have a command of the waterways matched by few other animals. These massive ungulates, weighing more than 2½ tons, are truly amphibious, seldom wandering more than a few kilometres from water.

In the water, the hippo controls its buoyancy at will by simply regulating the volume of air in its enormous body. Because its specific gravity is slightly greater than that of water, the retention of a little more air in the lungs causes the hippo to float, allowing it to swim when necessary. By expelling air from its lungs, it sinks slowly to the river bed, where it tiptoes with surprising grace.

Hippo may cover a considerable distance in the six minutes they can stay submerged. In doing so they perform a valuable ecological function: the underwater movements stirs up sediment at the bottom of the pool or river, making food and nutrients available to other organisms. Hippo also help to clear water channels.

Calves may be born in water or on land. Within five minutes of its birth, a hippo can walk, run and swim. It usually prefers the safety of submersion when suckling. At birth, the calf may weigh anything from 27-50 kilograms.

Hippo generally spend most of the day in the water, though they emerge to sun themselves when it is cold. In spite of a red 'sweat' which protects them from the effects of water, and ensures their skins do not dry out on land, hippo can suffer from sunburn.

Evolution has fashioned the hippo's eyes, ears and nose to project like periscopes from the large, flat face. The nostrils – little more than slits – close when the hippo submerges. The eyes and ears are often all one sees of hippo, but they occasionally expose their broad backs – providing convenient perches for various birds. Wagtails are attracted by the aquatic insects suddenly made available; oxpeckers and cattle egrets clean the massive backs of skin parasites.

At night, hippo leave the water to graze. Like white rhino, they have mouths ideally suited to eating soft, short grass. However, other plants, fallen fruits and occasionally aquatic vegetation are also part of their diet. They consume about 50 kg of vegetation a night.

Like all wild animals, hippo must be treated with respect, and never approached too closely, whether on land or in the water: they will defend themselves vigorously. Hippo are at their most dangerous when trapped away from water. They trample distinct paths to their feeding grounds, and if disturbed will use these routes to quickly return to the water. Humans have been killed because they were unlucky enough to be caught between a hippo and the water.

Hippo have few natural enemies. They even live in harmony with crocodiles, although these reptiles will attack calves. Lions will also tackle immature hippo.

Because hippo are largely free of predation, and because of their extensive diet, man has to intervene and control their populations by culling in national parks and game reserves. Over-population would put excessive grazing pressure on the river banks – particularly during droughts. This would make life even more hazardous for other species, and increase the chances of soil erosion.

Late in the afternoon, a procession of elephants with their young approaches the Caterpillar Pan in Hwange National Park to drink. Elephants have a complex social system, with family groups comprising a number of related cows with offspring, led by a matriarch. This structure provides security for the young. At puberty the bulls leave of their own accord, whereas cows remain with the family group.

Dwarfed by the immensity of the African savanna, a herd of Burchell's zebra thunders across the unfettered wilderness of eastern Caprivi.

Disturbed from the air, hippos rush for safety in the Letaba River, in the Kruger National Park.

A white rhino, characteristically preceded by her calf, in the Rob Ferreira Game Reserve. She displays her prominent horn – today, sadly, a symbol of the species' decline at the hands of poachers who are lured by big money. The two main uses have been as an aphrodisiac in the Far East (there is no scientific basis for the legend) and as dagger handles in the Arabian peninsula, where they have a substantial status value.

Mighty lord of all he surveys, a bull elephant emerges after bathing in the Zambezi's Kariba Dam.

An elephant calf in the Addo Elephant National Park. The resident elephant herd faces serious problems: increasing numbers and the static habitat have created a classic conservation dilemma. Unless the park is enlarged, translocation or culling will become unavoidable.

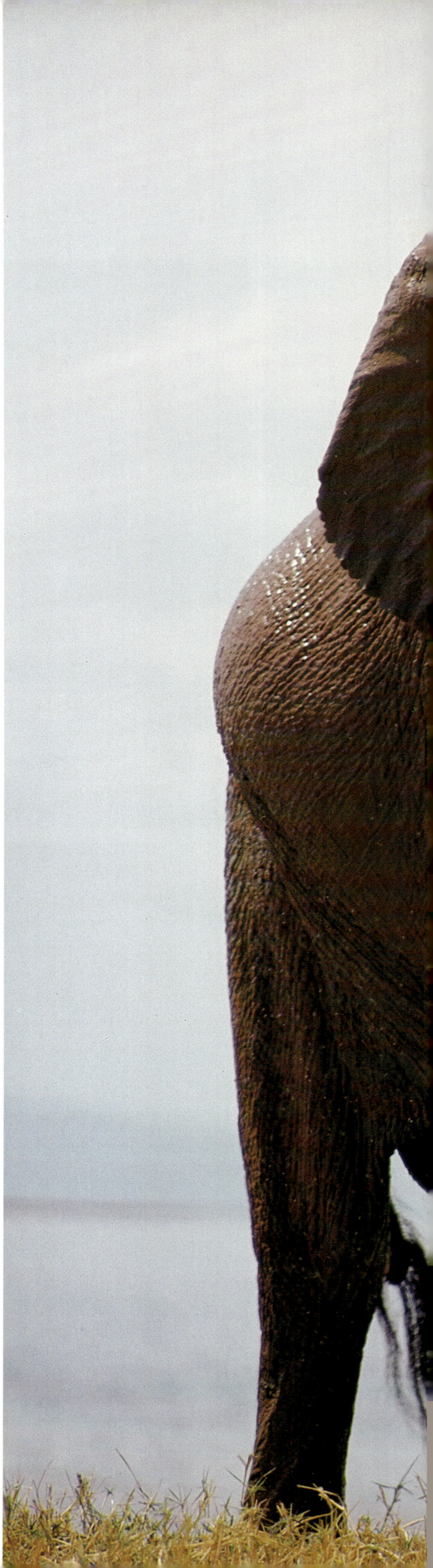

Right: A herd of Burchell's zebra has adapted to the karoo conditions in the Victoria West Nature Reserve.

Far Right: The bull giraffe (as shown here in the Mkuzi Game Reserve) has an average height of 5 metres, while a female is slightly shorter at about 4.5 metres. Although generally considered voiceless, giraffe grunt or snort when alarmed. Females make a whistling sound to call their young, and captive individuals have been known to moo gently when hungry.

An elephant at Mana Pools National Park reaches into an acacia for forage. Elephants are extremely destructive feeders, often pushing over a number of trees to expose succulent roots – to which they may give scant attention. This can have a severe impact on the vegetation in a fenced game reserve, necessitating population control by culling.

A group of hippo bask on the shallow banks of the Sabie River in the Kruger National Park. Although hippo usually spend the day submerged, they sometimes converge on river banks. They seldom stay out of the water for long, as their sensitive hides suffer from insect bites and sunburn.

Red-billed oxpeckers provide ubiquitous consort to hippos when on land, fulfilling the vital function of parasite control.

A pair of bull elephants head for the cool waters of the Letaba River. Kruger National Park.

Below, left: Being gregarious animals, Burchell's zebra are well adapted for life on the plains of Hwange National Park. Mares with young are at their most vulnerable when the foal suckles. Although foals begin grazing after 10 days, it takes a year before they are weaned. Disease, lack of food and predation claim half the season's young.

Below, right: They may look placid, but buffalo like these in the woodlands of the Kruger National Park have a reputation for being cunning and ill-tempered, especially if wounded. Their size daunts most predators, but lions may attack cows and their young.

Its yellow, black and red bill and crimson knees and feet give the Saddlebill stork a comical appearance. This bird may be seen singly or in pairs along the shores of Kariba, or in marshy areas throughout the region, stalking frogs, fish, crabs or shrimps with slow, deliberate strides.

Previous page:
Above: A herd of buffalo grazes peacefully in a grassy clearing, surrounded by the russet shades of mopani woodland, in the Hwange National Park.

Below: A herd of Burchell's zebra in the Moremi Wildlife Reserve. Zebra roam in huge numbers over the pristine wilderness that still covers large areas of northern Botswana.

Left:
White rhino exhibit their characteristic walking gait, with heavy heads close to the ground, as they graze through a forest of aloes in the Pietersburg Nature Reserve.

Below, left: Hippos crowd along the exposed banks of a reed bed in the Okavango Delta. These animals fulfil an important ecological function by opening up channels in rivers, which keep the water flowing, and stirring up sediments which provide food for fish and other creatures. However, like elephants, they have to be culled, as their impact on the habitat may be most destructive – especially during droughts.

Below: Although a number of giraffe may sometimes associate, they do not form large herds. Groups are generally made up of females with their young, but mixed or bachelor herds, as shown here in the Kruger bushveld, may also occur. Giraffe are not territorial animals, and bulls drift from one group to another.

A herd of buffalo makes its way across the Shingwedzi River in the Kruger National Park. Dry for some nine months of the year, the river is transformed in the rainy season to a fast-flowing ribbon of water which temporarily impedes the free movement of game.

A giraffe calf in Londolozi Game Reserve.

In the Chobe floodplain of northern Botswana, a family group of elephants wallow and splash in a mud pool. This behaviour has the advantage of destroying skin parasites, and enables the elephant to cool off in the blistering mid-day heat.

The open thornbush savanna plains of Etosha National Park provide an ideal habitat for the stately giraffe.

Buffalo graze peacefully in an autumnal setting of mopani bush in the Hwange National Park.

Hippo pathways lead to a herd among the reeds of the Okavango Delta. Seldom far from water during the day, at night hippo traverse well established pathways to grazing areas which may be up to a kilometre or more distant. If alarmed they return at speed down these familiar routes to the water.

Right: In water, hippo tend to bask close together, but an individual's sudden shifting can easily arouse an aggressive response from its neighbours.

Prolonged drought in the Chobe National Park has severed the waterway linking the Savuti channel and the perennial Linyanti River to the north. Here a large herd of hippo wallows in one of the few remaining pools of the Savuti, which is itself fast succumbing to evaporation in the searing heat.

Far left: An African spoonbill at its nesting colony in the Bluff Nature Reserve, Durban.

Left: Hippos trade threatening gestures as the sun sets over the Kruger's Letaba River.

A herd of elephant crosses the dry veld in the Kruger National Park. Because elephants eat 100-200 kg of vegetation daily, they have to traverse large areas in search of food. The bulk of their diet consists of leaves, bark, acacia seed pods, marula fruit and grasses. Consequently, their worn-down molars are replaced six times in their 50-60 year lifespan.

Hippos are voracious grazers, consuming about 50 kg of vegetation each night. Although their diet also includes aquatic vegetation and fallen fruits, their broad mouths are well adapted to eating soft, short grass.

A giraffe bull and cow follow their calf towards better browse in Swaziland's Mlilwane Wildlife Sanctuary. Patterns on a giraffe display a remarkable variation in shape and colour, varying from pastel shades to deep brown. Giraffe in thick bush may be sexed by their horns: in contrast to the bull, a female giraffe has prominent tufts of hair growing on the tips of her horns.

Far right: Black rhino once occurred in most parts of southern Africa where the habitat was suitable, but hunting, poaching and habitat destruction have so greatly reduced their numbers that the population is today probably no more than 2 000, and the species is regarded as endangered.

Young buffalo bulls spar in a mock contest.

Giraffe fill a unique ecological niche, as they browse on the top of the tree canopy, thus eliminating competition with other herbivores.

Far right: Her family group threatened, an elephant cow at Chobe National Park adopts a challenging attitude; ears spread wide, trunk held firmly forward, feet stamping, trumpeting and snorting loudly. This awesome spectacle, which usually puts intruders to flight, may be followed with a mock charge which stops short of the target. Sometimes, however, the mock charge may turn into a real attack.

Below: Perched among the reeds of the Okavango Delta, a brilliantly coloured Malachite kingfisher maintains constant vigil for the small fish, tadpoles and frogs on which it feeds.

A principal difference between white rhino and black rhino is the shape of the head. White rhino have long heads with broad mouths for grazing, whereas black rhino have bullet-shaped heads with a prehensile lip for browsing.

Far right: Burchell's zebra in the Victoria West Nature Reserve.

Stately, disciplined, the young elephants follow on the heels of the cows beside the waterhole at Goas, in the Etosha National Park. When on the move, eating or drinking, elephants keep in contact by making a loud rumbling noise in the throat. If one elephant detects danger, it stops rumbling, thereby warning the others.

Previous page: With thundering hooves, a herd of buffalo stampedes across the plain in the Matusadona National Park.

Fighting is less common among zebra stallions than in males of any other species, but might occur when a colt tries to steal a mare to begin a family of its own. These two Cape mountain zebra at De Hoop are exhibiting the classic attack pattern, which involves biting at the neck, throat and mouth and rearing up on the hind legs, lashing out with the front hooves.

During the dry winter months, elephants in their hundreds converge along the banks of the perennial Chobe River.

A white rhino bull grazes in the Umfolozi Game Reserve. Short grasses are its exclusive diet.

A giraffe is at its most vulnerable when, with front legs awkwardly splayed, head down, it stoops to drink. These animals usually visit waterholes in small groups, but drink individually while the others keep watch. The jugular vein, connecting the large heart to the brain, is lined with special valves to prevent blood from flooding to the head when the giraffe drinks.

Right: Showing signs of a recent mud wallow, a white rhinoceros and her calf stand warily on the grassy flats of the 1 500 ha Rolfontein Nature Reserve in the Orange River valley. This reserve was recently proclaimed to conserve a habitat where rhino and other species were formerly abundant.

Giant herds of buffalo still occur in the wilder regions of the subcontinent. Massed beneath the acacias of Mana Pools National Park, a herd wanders freely through one of the last stretches of unspoilt Zambezi floodplain. This conservation region has for several years been under the threat of the proposed Maputa Gorge Dam scheme which, if followed through, will spell ecological disaster not only to the prolific wildlife, but also to the unique forests of Acacia alba.

Below right: A herd of 'desert' elephants moves slowly along the dry Uniab River bed in Damaraland. These elephants have successfully adapted to a harsh semi-desert environment, where conditions force them to cover large distances in search of food and water. Poaching has taken a heavy toll in recent years, and protective measures have proved vital in maintaining the breeding viability of the remaining herds.

Overleaf: A lone tusker beside the Kariba Dam in the Bumi Hills region.

A Cape mountain zebra stallion stands guard in the Akkerendam Nature Reserve, a typical karoo habitat of this subspecies.

Hartmann's mountain zebra, a subspecies of the mountain zebra, occurs only in certain parts of Namibia, favouring the ecotone between coastal mountain and flats. The Skeleton Coast Park is one such typical habitat where, if surface water is inaccessible, zebra will dig for it with their hooves, sometimes excavating up to a metre deep.

Far right: The markings on the Cape mountain zebra (shown here in the De Hoop Nature Reserve) differ significantly from those of the common Burchell's zebra, with stripes covering the length of the legs, and not stretching across the belly. The Cape mountain zebra has larger ears, and a reddish-brown muzzle.

Baboons and Monkeys

Chacma baboons, vervet monkeys and Samango monkeys belong – with humans – to the order of primates. The most important features of this order are the large brain and higher intelligence. There are many different levels of intelligence within the primate order.

Most primates have adapted to an arboreal existence: their hands and feet have an opposable thumb, enabling them to grasp branches easily as they move through the trees. Sight and hearing, rather than smell, are their most important senses. Like humans, their binocular vision enables them to judge distances; they also have flat nails rather than claws.

Primates are further divided into anthropoids (more advanced species) such as humans, apes and monkeys, and prosimians such as bushbabies. Anthropoids have better-developed brains, and generally have flatter, more expressive faces. Prosimians favour the night hours, whereas their relatives are diurnal.

At one time, baboons were probably forest dwellers – living, eating and sleeping in trees, as many primates do today. Why they came out of the trees is a matter for conjecture, but it might have had something to do with gradual climatic changes which caused the forests to shrink, while the savanna plains increased in size. To survive, baboons had to adapt to a new environment.

The new environment was far more dangerous than the forests, where their only enemies were probably raptors and snakes. On the plains, they were faced with such enemies as lions, leopards, cheetahs, wild dogs and hyenas. They also had to look harder for their food: it had been plentiful in the forests, but was relatively sparse on the plains. To increase their chances of finding food, and to protect themselves from predators, baboons formed troops.

A baboon troop can number up to 50 males, females and young. Normally, it is led by a large adult male, but sometimes there may be two or more dominant males. As is the case with so many gregarious species, the females are the troop's backbone. They are born into the troop and spend their entire lives with one troop. Males, however, may leave and join other troops on maturity.

Among the most important duties of the males is the protection of the troop from predators. Their teeth are very effective weapons: the canines are long, pointed and deadly. A male's 'yawn' is actually a threatening gesture – one which is recognised by most predators. A leopard, for example, may find the baboon a formidable adversary, and will usually move off if several male baboons advance on it.

From time to time the dominant male is challenged by another member of the troop, and has to defend his position. The ensuing fight can be vicious, and occasionally the wounds inflicted can lead to death. Battles for dominance are a way of ensuring that the most capable male leads the troop.

There are many other social rules and gestures which keep the baboon troop together and help members communicate with one another. 'Body language' – the way they yawn, grin or smack their lips – is an important means of communication.

Grooming, too, fulfils more than just a physical need. Ostensibly, the groomer is only removing bits of dead skin, dirt and any insects from another baboon's hair. However, the grooming also creates a bond between troop members and strengthens links between individuals.

Young baboons have to be initiated into the rules of the troop, and the quickest way of learning is through play. They spend most of the day at play – usually with others, but sometimes on their own. They run, chase and wrestle one another, and if the play becomes too rough, an adult baboon will put a stop to it. These games enable the animal to test its own strength, and that of other baboons.

Baboons are found in a number of habitats. They occur on plains, koppies and mountains throughout southern Africa (with the exception of the western desert regions).

On the plains, baboons spend the nights in trees. They do not always return to the same sleeping site each night, as their daily foraging can take them over a range of several kilometres. In fact, they may have several sleeping sites within their home range.

On mountains, baboons favour cliff faces for their nightly shelter. As the leopard is their most feared enemy, they seek refuges which are inaccessible to the big cat, and seldom leave the rocky crevices until daybreak.

Although baboons are mainly vegetarian, they are true omnivores, eating wild fruits, berries, leaves, grasses, roots, bulbs, tubers, insects, scorpions, centipedes, lizards, small birds and even small or young mammals. These primates can be extremely dangerous, and should never be fed by hand. They have caused serious injuries to people, and have been known to kill children.

There are many similarities between a troop of chacma baboons and a troop of vervet monkeys. Both species have strong social bonds which unite the troop. A vervet monkey troop consists of 20 or more individuals, led by a dominant male – usually the biggest in the troop. He determines the troop's territory and guides it in the search for food. While the rest of the troop forages, the dominant male stations himself in a tall tree to watch for predators.

The leader is intolerant of cheeky males, and will soon put them to flight. However, he is more tolerant of females and infants, allowing them to crawl about him and even eat his food. If two males fight, he acts as 'referee', intervening only if the battle becomes too serious.

Females are also ranked within the troop. The dominant female associates with the male leader. She seldom has to assert her rank with other females, as they quickly learn their place within the hierarchy. The females are responsible for disciplining the juveniles; a sharp nip at the base of their tails is an effective reprimand.

Vervet monkeys have a complex social life with many rules which have to be learnt by the young monkeys. As the infant becomes accustomed to occasional separation from its mother, it learns to play with the other young monkeys. Play is important for several reasons: it teaches the monkey about the habits and idiosyncrasies of other troop members and sharpens its agility – vital in its arboreal existence.

Males are allowed to stay in the troop until they are about four years old, at which time the top-ranking male expels them. Females remain with the troop, and are regarded as adults at the end of their second year. Status plays an important role in the monkey troop. For example, the female offspring of a high-ranking female is ranked just below her mother.

Young females commonly 'babysit' infants; this is known as 'aunting'. It is possible that the practice teaches them how to care for infants – in much the same way as young girls learn by 'playing mother'.

Grooming is as important among vervet monkeys as it is among chacma baboons. It creates a bond between monkeys as individuals 'ask' for attention by lying across a branch in front of another monkey, or by standing sideways on to the one they would like to groom them. Grooming also has a practical function: it keeps their fur in good condition.

Vervet monkeys have, where necessary, adapted to life alongside Man. Some are even at home well within an urban environment. Given the opportunity, they will raid a picnic table or an open kitchen. They can wreak havoc in vegetable gardens, uprooting carrots and beetroot. They have been known to destroy large quantities of fruit, and can ruin a crop of sugar cane. Their normal diet includes leaves, fruit, flowers and nectar, and they occasionally chew bark.

Vervet monkeys are not the only monkey species found in southern Africa, but they are certainly the most common, and their habits are fairly well documented. Samango ('blue') monkeys are more secretive, and not much is known about them. Vervet monkeys live in a variety of habitats, from forest to light savanna woodland, while samango monkeys prefer indigenous forest, and never move far from water. Sometimes, however, the two species are found in the same region.

Vervet monkeys are found from the south Cape along the east coast to Natal, in the eastern and northern Transvaal, Mozambique, Zimbabwe, the eastern and northern parts of Botswana, by the banks of the Orange river in the north-west Cape and northern Namibia. Samango monkeys occur from the Tsitsikamma forests up the coast to Natal. They are also found in the Kruger National Park, and in parts of Mozambique.

Two baboon mothers, freed for a few moments from the attentions of their infants.

Baboons spend most of the day on the ground, foraging for bulbs, tubers, grasses, small birds or lizards, and occasionally insects. At night the troop seeks the safety of closely grouped trees, not venturing down until daybreak.

A dominant male Chacma baboon yawns, signifying not boredom or fatigue, but a warning to any potential challenger to his authority.

For baboons, grooming is a daily ritual.

An infant baboon in the Kruger National Park.

Overleaf: A close bond is formed between a baboon mother and her offspring. Until her young is about eight weeks old, she grooms, feeds and carries it about with her, not letting other females pick it up until it is able to walk. Thereafter she tolerates attention towards her young from other members of the troop.

Grooming plays an important role in a Chacma baboon's social life, strengthening relationships between troop members. The instinct also has practical advantages: ticks, dead skin and dirt are removed in the process.

After a day's activity, a mother and her offspring relax in the late afternoon sun. Although at ease, the adult is always alert for danger, and at a moment's notice will grab her young and make off. Kruger National Park.

The vervet monkey is distributed throughout the African savanna south of the Sahara. It is a gregarious, diurnal species, usually seen in troops with a hierarchical social structure dominated by a male or oligarchy of males. As with some other species, a yawn, as shown here, indicates a warning. Bonding is strong within a vervet monkey troop, and at night, groups of individuals sleep together according to rank. Grooming enhances relationships, and is an important mid-morning activity after foraging.

In a tree beside the Letaba River, in the Kruger National Park, a vervet monkey with infant feeds on fresh leaves.

Play among young baboons helps them to learn the rules of behaviour within the troop. Baboons give birth to a single infant.

Groomer and groomed.

A Chacma baboon rips the husk from the fruit of a 'sausage tree' in the Kruger National Park.

Antelope

A glance at the fossil record shows that antelope have a long history – a history going back 65 million years, to a time when they were small hornless creatures. From these remote ancestors have evolved many distinct species, of which 33 species occur in southern Africa. The word antelope, meaning 'bright-eyed', refers more to gazelles – those medium-sized, slender, graceful creatures – than to buck in general. All, however, belong to the Bovid family, which includes giraffe, deer, camels, pigs and hippo.

Bovids are characterised by their hooves and four-chambered stomachs. They lack upper incisors, and chew the cud. Males have unbranched, bone-filled horns which are not shed. In some species the females, too, have horns. All bovids have an even number of toes.

Evolution has adapted antelope to every type of habitat Africa offers: from swamps to deserts, mountains, forests and grasslands. But the process of changing has not stopped. Species are still being modified to survive in changing environments. Unfortunately, the process of altering has been artificially speeded up by Man. Sometimes his changes are too drastic and too fast for nature to counter-balance. Indiscriminate hunting, poaching, closure of migration routes, ill-advised veld burning, bush encroachment on grasslands, over-grazing, afforestation, poor watershed management and the long-term effect of pesticides are among the ways in which Man degrades the natural environment. The results can be devastating: entire species become extinct.

The dwindling numbers of roan antelope are an example of how habitat destruction can push a species into decline. Densely populated rural areas and expanding farmlands have encroached on the lightly wooded countryside preferred by roan antelope. As a consequence, they have been driven into unfavourable habitats. Despite efforts to maintain viable herds, these creatures remain exceptionally vulnerable. In South Africa, a remnant population in the Kruger National Park have to be given special treatment, as they are prone to anthrax – a disease which could wipe them out. Fortunately, the species is more stable in Zimbabwe and northern Botswana.

While roan antelope favour specific habitats, some species – such as the grey duiker – have adapted to a wide variety of environments in southern Africa. These are probably the most successful of all Africa's antelope, even though their enemies include almost all the carnivores as well as eagles, owls and leguaans. Their survival owes something to their solitary nocturnal habits, and to their diet. Grey duiker are primarily browsers, but they also dig up roots and tubers, and eat bark, stems, berries and flowers. They have even been known to eat mopani worms. Being independent of water, they avoid the risks faced by other antelope, which have to drink daily. As a result, their chance of surviving habitat destruction is far greater than that of roan antelope.

Steenbok, too, have developed a flexibile diet which allows them to live all over southern Africa (except on the sub-tropical eastern seaboard). These little buck are both grazers and browsers, eating grass, leaves, young shoots, roots and tubers. Like grey duikers, they do not appear to drink, as they obtain enough liquid from their food. During dry seasons, steenbok can survive on a diet as meagre as fallen leaves. Although they prefer open grassy or lightly wooded country, steenbok have managed to survive close to cities – unlike many other animals.

Wildebeest, hartebeest, tsessebe, bontebok and blesbok are open grassland species, while springbok keep to dry plains. These are among the fleetest of the antelope. Tsessebe, regarded as the fastest antelope over long distances, have considerable stamina, and can keep up their seemingly awkward bouncing gallop for a long time. Both tsessebe and red hartebeest have been clocked at speeds of more than 60 km/h, while springbok can achieve 90 km/h in short bursts, making leaps of 15 m or more.

While these antelope have adapted to the plains, gemsbok prefer the desert and semi-desert regions of Namibia and the Kalahari. They have developed special ways of surviving in this harsh environment, feeding on grass and fruit (especially wild melons and cucumbers) as well as the bulbs of succulents. They extract most of their liquid requirements from the plants they eat. If water is available they will drink regularly, but they can go for long periods without drinking.

Although most antelope drink frequently when water is available, many species control their water intake. Tsessebe need drink only once a month, while grey duiker, kudu and roan can last two to three days without water.

For a gemsbok, the instinctive need to conserve moisture may rule out flight when danger threatens. Its flight would lead to moisture loss – and in the desert this loss could be serious. So instead, the gemsbok will often turn to fight. To this end it is armed with long, rapier-like horns of such formidability that even lions have learned to be wary. Relatively few gemsbok end up as victims of a hunt.

Perhaps the most significant adjustment gemsbok have made to their environment is their means of temperature control. So successfully do they retain and use body moisture that urination is frequently a cursory affair!

Gemsbok have a built-in 'air conditioner' to cool blood flowing to the brain. During the day, as the gemsbok's body temperature rises, the blood is fed through a fine network of blood vessels in the nose, where it is cooled by the air as the gemsbok breathes.

Swamp antelope, such as sitatunga, lechwe and puku (found in the Okavango and Linyanti swamps), spend most of their time in water between five and 50 cm deep. To walk over slippery vegetation, mud and silt, the sitatunga has evolved special hooves. They are elongated and splayed, spreading the animal's weight over a large area and enabling it to walk with ease in its waterlogged environment.

Another adaptation sitatunga have made to their semi-aquatic environment is an unusual way of hiding when danger threatens. These antelope are good swimmers, and if alarmed will head for deeper water. They then sink until only their nostrils and eyes protrude above the surface. They remain like that until the danger passes.

Sitatunga are shy and rarely seen. They live alone or in pairs, moving about during the day and night. They eat mostly swamp grasses, but also browse on leaves, twigs and the fruit of semi-aquatic plants.

Lechwe do not belong to the same family as sitatunga: instead, they are related to other water-loving buck such as puku, waterbuck and southern reedbuck. Yet like sitatunga, they have long, narrow hooves, they can swim well, and they take refuge in water. However, unlike sitatunga, lechwe and puku are gregarious, forming herds of up to 30 members.

Each antelope species has come to terms with its particular habitat in unique ways. For example, dense thicket and forest browsers, such as nyala and bushbuck, have assumed the tawny dappled shadings of the glades they haunt.

Klipspringers, which live on rocky outcrops and koppies, have narrow cylindrical hoofs – analagous, perhaps, to the hard tips of a ballerina's shoes – to give them purchase as they leap about the rocks. Their hair is brittle and springy as a protection against sharp rock edges. No other antelope has this sort of coat. While many species have adapted to year-long survival in their habitats, others migrate in search of better grazing as the seasons change.

In the past, antelope such as springbok, eland, gemsbok and wildebeest were able to undertake annual migrations to find fresh grazing and water, or to avoid parasites. Often their survival depended on mobility, but today – throughout Africa – fences curtail their freedom.

Gone are the great springbok migrations of the last century, when residents in Karoo towns clubbed animals to death as they streamed past their stoeps. Such were their numbers that when they came up against a wall, those in front would be crushed to death until a ramp of carcasses overcame the obstacle.

Explorer George Gordon Cumming tells how he was woken one morning to find the ground 'covered with a dense living mass of springbok, marching slowly and steadily along, extending from an opening in a long range of hills on the west, through which they continued pouring, like the flood of some great river, to a ridge about a mile to the north-east, over which they disappeared.

'The breadth of the ground they covered might have been somewhere about half a mile. I stood upon the fore chest of my waggon for nearly two hours, lost in wonder at the novel and wonderful scene which was passing before me, and had some difficulty in convincing myself that it was reality which I beheld, and not the wild and exaggerated picture of a hunter's dream.' Some researchers estimate those enormous herds of springbok could have numbered in the millions. In one instance, a trek in Namaqualand resulted in a great number of springbok perishing when they reached the sea and drank the salt water. Carcasses were said to litter the shore for 50 kilometres.

Today, the southern continent is criss-crossed with fences, making migration impossible in many areas. Even in remote parts of the Kalahari, game-proof fences are creating havoc by blocking the routes of antelope responding to strong migratory instincts.

One of the most contentious barriers – erected to prevent the spread of foot and mouth disease – cuts across the migratory path of blue wildebeest, which migrate from the central Kalahari to forage along Botswana's lakes and rivers. Each year the fence takes a massive death toll – a toll made all the more tragic because little is known about the transference of foot and mouth disease between wild and domestic animals. The phenomenon has never been demonstrated.

Nevertheless, fences exist, and more are planned. The result has been a disastrous mortality rate of hundreds of thousands since the fence was erected in the early 1950s.

Under natural conditions, the wildebeest filter through the veld in relatively small groups. However, fences focus their migratory movement along a fixed route. Unable to go through the fence, the animals drift eastwards alongside it, building up in numbers and consuming and trampling what little fodder is available. Instead of dispersing throughout the arc of riverine habitat formerly available to them, the entire population is concentrated by the fences into one small area. Here they mass, trapped by human habitation to the north and south, and by their need for water. When all the food has gone, many starve to death. This is a tragic illustration of Man's unthinking interference with nature.

Wildebeest are not the only animals whose migratory patterns are upset by human intervention. For example, eland living in the Drakensberg region are forced to move to lower altitudes during winter. Here again, their natural movements are restricted by farm fences. In the Cape, the steep decline of the reedbuck population is also blamed on farm fences.

The survival of a species depends not only on its habitat, but also on its reproduction rate. In all the large antelope species, reproduction involves ritualistic behaviour to ensure that the best genes are passed on.

Most antelope form territories marked either by dung deposits, urine scenting, gland secretion or earth mounds (dug with the horns). In species that are not territorial, breeding is carried out by a dominant male. Territoriality, or dominant male breeding, is characterised by highly evolved codes with endless variations, depending on the species and their habitats.

The size of a territory varies from a roan bull's 50 hectares to the 5 hectares defended by a pair of dik-diks. Some antelope are particularly pugnacious: waterbuck, grey duiker, oribi and kudu do not tolerate border violations.

Antelope are generally gregarious, forming herds or troops numbering 6-12 individuals. Lechwe herds can number as many as 100 individuals. Solitary kudu, sable or wildebeest bulls are usually sickly, or outcasts as a result of a lost territorial battle. Sociability is not a trait of the smaller species: duiker, dik-dik, steenbok and klipspringer are usually found singly, or in pairs.

In the larger antelope species, harems made up of females and young are presided over by a bull or – in the case of kudu and eland – two bulls. Nyala, too, conform to this pattern, as the troop's leader drives off young bulls which subsequently form separate troops. Female waterbuck are less restricted, and are free to range through the territories of a number of bulls.

After the mating season, the bulls of some species form bachelor herds. In the case of territorial species, these are 'psychological castrates' – without territories, and therefore unable to breed. In the constant battle for supremacy, however, a male from a bachelor pool could unseat an established territorial male.

Among kudu, for example, old bulls defend their territory during the mating season, sometimes with fatal consequences. Yet after the breeding season, adult males will desert the females and young to form bachelor herds until the following season.

Reproductive cycles in antelope are similar to those of other mammals. Most species breed once a year, giving birth to a single calf at the most advantageous time of the year – usually the rainy season. Smaller species, especially forest dwellers, produce young without any seasonal preference.

Sexual maturity in antelope is related to their size. Smaller species have a life expectancy ranging from three to four years and are sexually active at 12 months. Larger antelope, who live for about a decade, reach maturity in their second or third year.

Mating seasons are determined largely by such environmental factors as length of daylight, temperature and food availability. These stimulate the pituitary gland, heralding the onset of oestrus. The females now exude a characteristic scent.

This scent heralds the start of vigorous courtship rituals among males. Courtship is an elaborate combination of attraction, appeasement and reassurance behaviour. Common to both sexes among large antelope is an instinctive aversion to closeness. However, to produce offspring, they must alter their attitude to physical contact. In addition, the male must convince the female that the highly excited creature prancing before her is not dangerous. And she must stand still long enough for the mating to be consummated.

If all mature animals were to produce young, there would inevitably be more deformed individuals, and the amount of sub-standard genetic material would increase. This would clearly not be in the long-term interests of the species.

Each species has what is known as a gene pool. In this pool are all the genetic variations which keep that species healthy. Without the rivalry rituals, the gene pool could become infected with harmful genes: these could become dominant, placing the entire species at risk. That risk can be minimised if the parents' contribution to the next generation is of the highest quality. Courtship, with its horn clashing, fencing and parrying, bellowing and roaring, is simply nature's way of deciding which animals are best qualified to contribute to the gene pool.

In the highly elaborate courtship rituals, aggressive displays are more frequent than actual combat. When contact is made, the horns are usually locked together, so that little damage occurs – though there is a great deal of noise. But at the same time there is always an element of risk.

Among kudu, courtship fights can be fierce and sometimes deadly. Should their horns entwine and lock in such a way that the combatants are unable to pull free, both could succumb to starvation or predators. Sable antelope and roan antelope are also great fighters in the mating season. In a typical encounter, the bulls of these species will sink to their knees, sweeping their horns furiously from side to side. Bulls are sometimes fatally wounded during these clashes, but most 'fights' are simply harmless displays of aggression. For obvious reasons, it would not be in the interests of the species for courtship deaths to occur on a large scale.

Impala sometimes fight to the death, but such incidents are the exception rather than the rule. Grey rhebok also indulge in fierce but harmless mock fights, but will sometimes attack other species: they have been known to kill mountain reedbuck and even sheep and goats.

Because the males of many species spend so much time and effort establishing territories, dominance, and the right to mate, in the end the strongest, most determined, cunning and quickest win. As the mating season peaks, courtship ceremonies intensify. Obsessed with their power struggles, males may not eat for days – oblivious to everything but the battle for supremacy. This further separates the weak from the strong.

Young springbok rams, for example, have a high mortality rate attributable to stress or starvation. This is hardly surprising, since territorial rams spend a great deal of time defending their territory and courting ewes. Particularly vulnerable are young, inexperienced territorial rams, which are relegated to less desirable grazing sites – with tragic results.

Springbok do not have reserves of fat, so they need to feed regularly. If deprived of food for any length of time, they may die of starvation or stress.

Courtship ends once fertilisation takes place. The gestation period varies from four and ten months, depending on the species. When the birth is imminent, the females adopt various strategies to ensure their own survival and that of their offspring in those vulnerable hours during and after the birth. Probably the most common tactic is for the female to leave the herd and hide shortly before the birth. Because she cannot afford to expose herself to danger for any length of time, labour is brief. Most births occur at midday, the hottest time of the day, when predators are most likely to be inactive.

The behaviour of impala ewes is typical of this pattern; they give birth at midday in thick bush. Once the calf is born, the ewe bites through the umbilical cord, licks her offspring, and eats the afterbirth. The calf stands almost immediately, and is playing within a few hours. Mother and offspring are able to rejoin the herd within a short time. If the calf does not respond normally, she keeps it hidden for a few days before venturing out to find the herd.

This birth behaviour is common to grey duiker, oribi, nyala and gemsbok, although the larger species may hide their young for two to three weeks before rejoining the herd. In these instances the mother usually remains near her young, though waterbuck conceal their calves in tall grass or undergrowth and visit them only to allow the youngster to suckle. For the rest of the time, she stays well away for fear of attracting the attention of predators to her calf. She eats her calf's dung for the same reason. Some of the smaller antelope, such as steenbok, may give birth in disused aardvark holes, which become nurseries. Steenbok do not leave the nursery until they are almost full-grown.

Wildebeest cows often give birth in groups of 20 or more, the other females 'standing guard' against jackals and hyenas. The cow usually lies on her side to give birth, and as soon as it is over she jumps up to lick the calf and remove the foetal sac. The afterbirth is left for scavengers. Within 3-7 minutes, the calf is on its feet and ready to suckle.

Mothers of newborn calves form 'nursery' groups, but the young wildebeest always remain with their mothers. If a calf can survive the first few days, its chances of escaping attack by predators improve significantly, because it will already be able to run as fast as the adults in the herd. Unlike many young animals, a wildebeest calf can graze within 12 days, another factor which increases its chances of survival.

Nursery groups are also formed by other species, such as eland and impala. Among eland, the grouping of females and young is probably a defence against predators, as the calves are less noticeable, and can be protected by the adults. Eland females fight bravely to protect their young, and have been known to join forces to chase off packs of hyenas and wild dogs. There is even a record of a large group of cows attacking a lioness – a predator few animals will challenge.

Within an impala 'crêche', a calf will spend more of its time with its contemporaries than with its mother. The young may even be left unattended when the females are feeding, or in search of water. However, the adults will never stray so far that they are unable to return quickly at the slightest sign of danger.

Mothers commonly lick their offspring after the birth, thereby reinforcing the maternal bond. This is especially important because the mother has to be able to recognise her calf – and *vice versa* – in the confusion of the herd. Lactation lasts for about as long as pregnancy, and during this time the mother and young are together constantly. With the end of lactation comes a new breeding cycle.

On the grassy plains, at waterholes or in the forests, antelope are faced with a large formidable array of enemies. Lions, leopards, cheetahs, hyenas and hunting dogs threaten the larger antelope – in particular their calves – while genet, serval and jackals hunt the smaller antelope. Pythons and birds of prey also threaten the young.

In the endless struggle to survive, antelope have developed many ways of sensing danger and minimising the predator's threat. A herbivore's head is designed to detect danger. It has a large nose which is linked internally to a complex honeycomb of scent-sensitive membrane capable of detecting molecules in the air and identifying their source with a sensitivity far superior to that of humans. The ears are large and well developed and the eyesight is good, with a wide field of vision. Dik-diks also have good rear vision, due to their large, bulging eyes.

Camouflage is another of the antelope's resources. Thanks to their colouring, kudu in a thicket, or nyala in a glade, are particularly well hidden. 'Flash' colouring is another effective defence mechanism. Bushbuck, for example, have conspicuously white tails which they flash intermittently at pursuing carnivores. The predator has a momentarily clear signal on which to fix, but it vanishes in the next instant. This probably disorientates the hunter, much as a light flashed into someone's eyes on a dark night would cause temporary blindness.

'Going to ground' is a form of passive defence common to smaller antelope, which are able to fit into suitable crevices such as abandoned aardvark holes. Reedbuck, steenbok and puku, for example, lie flat and inert when danger threatens, exposing themselves only at the last moment to flee in a plunging, zig-zag dash.

The springbok employs an unusual evasive tactic known as 'pronking' – an awkward, stiff-legged jump. On occasion the jumps are too exuberant, however: springbok have been known to break their backs. Oribi exhibit a similar form of behaviour. When alarmed, they start off briskly and then begin a curious, stiff-legged spring straight into the air. Impala are also good jumpers, leaping up to 3 m high in flight – possibly to orientate themselves.

When attacked, sitatunga, lechwe and waterbuck take to water. If the water is too shallow to deter the predator, waterbuck will defend themselves with hooves and horns; otherwise they take refuge among reeds, submerged up to the nose. Legend has it that waterbuck give off a crocodile-repelling odour. Judging by the way they plunge unheedingly into crocodile-infested water, it would seem that they, too, believe the legend. However, records show that crocodiles will attack waterbuck as readily as they do other antelope.

As a last resort, some of the larger antelope – such as nyala, wildebeest, gemsbok, sable and roan – may stand and fight a predator. Even the smaller bushbuck puts up a valiant display.

Wildebeest defend themselves vigorously, forming a laager when threatened. Sable and roan will back into a bush and take on a lion if necessary. When under attack, bulls go down on their knees, scything with their deadly horns.

But there are other, more insidious threats to an animal's survivial – threats too small to be detected by the naked eye. These microscopic parasites are as much a part of an ecosystem as lions or hyenas, with a similar influence on the animals: they help to ensure that only the most healthy and vigorous survive.

At first, disease may seem an aberration in the wild, but in fact it is a natural part of an ecosystem. Conservation authorities are not worried about endemic diseases, but they keep a sharp lookout for 'new' or 'foreign' diseases. An epidemic could have alarming results. In biological time, an ecosystem may adjust to almost any disease, but in today's beleaguered sanctuaries, that time does not exist.

A classic example of a 'foreign invasion' is the highly contagious rinderpest virus. The disease swept through southern Africa at the turn of the century, all but wiping out buffalo, kudu and eland in the Lowveld.

One of the most infamous epidemics of recent times is anthrax. During the Second World War, experiments were carried out with the intention of using anthrax as a biological weapon. Even the most hardened scientists shuddered at the consequences of it being unleashed. Anthrax spores can lie dormant for up to 60 years and still be extremely potent.

Anthrax is highly contagious. It is spread by vultures and blowflies from the flesh and blood of anthrax victims. Kudu are particularly vulnerable. They may pick up spores at waterholes frequented by vultures carrying the disease, or from fly-blown foliage. Once infected, kudu die within one or two days.

In the 1960s, anthrax almost exterminated roan antelope in the Kruger National Park. Today, the Park spends about $20 000 each year on immunising these animals. At present, they are the only species being immunised.

Whether human intervention is necessary to ensure the survival of an antelope species, or any other wild animal, is not always obvious. This planet has long thrived on diversity and flexibility, yet the last 50 years have seen vast land tracts planted with monocultures, and thousands of species rendered extinct. The long-term impact of these developments on the biosphere is unknown, but one important truth has been recognised: genetic strains in certain crops are weakening, and the gene pools that could have been of benefit no longer exist. To destroy the earth's diversity of species is to endanger Man's survival.

Some antelope are Africa's original 'cattle'. If the situation is looked at pragmatically, they could be farmed and 'harvested' for their high-grade protein – part of the answer to the continent's crying need for food. On an equally practical level, wildlife is a tourist attraction, drawing increasing numbers of visitors to southern Africa each year. If and when the dust raised by political in-fighting finally settles on the sub-continent, the various countries will be able to offer wildlife safaris without precedent. With careful planning and foresight, the wild realms of Namibia, Botswana, Zimbabwe, South Africa and Mozambique could become huge generators of foreign exchange. Nowhere else in Africa is offered such a spectrum of habitats – from desert to swamp, savanna grasslands to forest.

Conserving wildlife is also a matter of conscience: animals have a right to live. Man has a moral obligation to uphold that right, and a responsibility to ensure that his rich natural heritage is not destroyed for future generations. If this responsibility had been acknowledged a little earlier, blue buck and quagga – both endemic species – would not be extinct.

The last bluebuck was shot near Swellendam in 1791; the last quagga died on the stone floor of its cage in the Koninglijk Zoologisch Genootschap in Amsterdam on 12 August 1883. Because this oddly marked animal had been a great attraction, the zoo wrote to the authorities in the Cape Colony and requested another specimen. By then, however, it was too late. It is not known when quagga became extinct in the wild. A herd of about 50 was known to exist in the Colesberg district up to 1873. It is thought that the last of the wild quagga in the Free State had died by the end of the 1870's.

The quagga was a zebra-like animal with stripes on its head and neck only. The back and rump were brownish, and the belly and legs were coloured light cream. It is a matter of scientific contention as to whether the quagga was a separate species or a sub-species of Burchell's zebra.

Quagga lived in the Karoo and Free State, but were not found north of the Vaal River. Burchell's zebra occurred in the Transvaal and Free State, but not south of the Orange River. It is unlikely that two different species of zebra could exist in such a small area as the Free State grasslands without one species becoming dominant and edging out the weaker. If the quagga was a sub-species, however, the two could have mixed without competition.

Some wildlife researchers have advanced a further theory in support of the quagga as a sub-species. In some instances the stripes on the hindquarters of Burchell's zebra in South Africa are so faded that they can hardly be seen. It is possible that the markings on zebra in the south are less distinct than those found in central Africa. Possibly the quagga had evolved to the stage where the stripes on the body had disappeared altogether.

Hunters of the last century also took a heavy toll of bontebok – another antelope unique to the Cape. Bontebok are often confused with the blesbok, a wide-ranging antelope to which they are closely related. They are similar in size, shape and colouring, but bontebok are larger, darker and more richly coloured. They have a white band running from the base of the horns to the nose; in the blesbok this band is broken at the eyes. Bontebok have a pure white patch on the rump, whereas this mark is pale brown on the blesbok. Bontebok occur only in the southwestern Cape, while blesbok are found on the grasslands of the interior.

In 1837, a farmer named Alexander van der Byle had the foresight to protect 37 bontebok in a special camp on his farm. At the turn of the century, their population numbered 300, but by 1930 it had dwindled to twenty.

To prevent the herd from further decline, 17 bontebok were transferred to a small reserve between Bredasdorp and Agulhas which became known as the Bontebok National Park. The area was far from ideal, so 84 bontebok were translocated to a new park near Swellendam in 1960. By the mid-1980's, the population numbered more than 360, and bontebok had been transferred to other Cape reserves.

Despite tough legislation in most southern African states, many antelope species are endangered. The situation is worsened by the fact that populations are isolated in various game reserves, with no chance of revitalising their gene pools. Humanity's territorial demands pose another threat: species considered 'safe' today could become endangered unless the sanctity of their habitat is assured.

While a species may be considered rare or endangered in one southern African country, it may have a healthy population across the border. Roan antelope, tsessebe and sable are vulnerable in South Africa, while Lichtenstein's hartebeest and oribi are declining in Zimbabwe. Swamp-dwellers (puku in particular) and lechwe are decreasing in number. Among the smaller species, the status of klipspringer, reedbuck and oribi is unsatisfactory.

Invariably, the problem concerns inadequate living space for these species: the more limited the distribution of an antelope, and the more specialised its habitat requirements, the greater the peril to its survival.

Although more research is needed before the problems that beset the antelope species and their habitats are overcome, considerable headway has been made in the last two decades. Capture and immobilisation techniques have been perfected, and the treatment of shock in newly caught animals is better understood. The feeding of wild animals in captivity is more successful. Even more important: the creation of new parks and reserves has meant the re-introduction of antelope into areas where they had long been extinct.

Previous Page: Among the best-known and most common antelope in southern Africa, the gregarious impala often congregates in herds of 50-100 during the dry months. Ewes form herds which include the season's young, whereas the rams gather in bachelor groups until the rut, when breeding territories are established.

Left: The Damara dik-dik is the only one of the six subspecies in the genus found in southern Africa, where it occurs in north-western Namibia. This tiny, graceful antelope is usually seen singly, in pairs or in small family groups. It inhabits dry, stony thickets and dense woodland, where it browses on low undergrowth.

Bushbuck are secretive and seldom seen, preferring thick bush or forest glades to the savanna plains. They seldom stray far from water. Mainly nocturnal, they tend to rest in the heat of the day, appearing in the late afternoon or early morning to browse on acacia pods and bushy shrubs. Of the three subspecies found in the region, the Chobe bushbuck (a ewe and ram are pictured below) have a lighter, more rufous appearance, with distinctive white spots and stripes on flank and belly, than their common cousin (above left), as portrayed here in the Kruger National Park.

A flight of greater flamingoes approaches a pan in the Lichtenburg district.

Opposite: Set apart from its common cousin by a distinctive shading between eyes and nostrils, the black-faced impala, shown here in the Etosha National Park, has a range confined to parts of northern Namibia and southern Angola. Its status is today extremely precarious due to civil disturbance throughout the region.

Right: A puku ewe amid lush grasslands in Chobe's floodplain. This species of antelope occupies a niche between lechwe, which favour reedbeds and open waterways, and waterbuck, which prefer riverine bush.

Below: With their high shoulders and sloping backs, tsessebe may look awkward animals, but they can sustain their lumbering gallop for long distances. Kazume Pan National Park.

Left: A gemsbok gallops across the vast expanse of southern Namib desert. Under such harsh conditions, where fresh water sources are few and far between, gemsbok obtain liquid nourishment from grasses, melons or succulent roots and bulbs, which they dig out with their hooves.

Top: For most of the year, the Kalahari Gemsbok National Park provides scant grazing for the blue wildebeest. Yet except under the most extreme conditions of drought, this ungulate is able to flourish.

Above: At the western limits of the Soutpansberg, a stretch of Kalahari sands provides a unique Transvaal location for gemsbok. The Langjan Nature Reserve was established to conserve the only naturally occurring population of this species in the entire region.

Far left: Roan antelope in mopani woodland, Hwange National Park. Lightly wooded savanna, with tall grassy regions and readily available water, are essential requirements of roan antelope. Today this species is threatened, as its very specialised habitats have been destroyed by development, and the prevailing drought over much of its territory has reduced the grass cover which is known to provide about 90 per cent of its diet.

Left: A young sitatunga in the papyrus swamps beside the Linyanti River in northern Botswana. The sitatunga is adapted for a wholly aquatic existence. Long splayed hooves enable it to move with ease through the swamps and marshes. When surprised in water, the antelope will completely submerge its body, with only the tip of the nose showing above the surface.

White-faced whistling ducks are common residents of inland waterways and lakes, where they can often be seen wading in the shallows.

Although predominantly nocturnal, Sharpe's grysbok also forage during the early morning or late afternoon. They usually prefer habitats of low-growing shrubs or high grasses, but have been found among stands of mopani. They are slightly smaller than the Cape grysbok, with a richer, reddish-brown colouring.

Young Springbok males indulge in playful sparring. Although adult males lock horns and 'wrestle' during the mating season, one combatant will usually withdraw before any serious injuries are inflicted.

Springbok once roamed in their millions throughout the drier regions of southern Africa, but the advent of the white man – with his guns and fences – led to the slaughter of countless animals. Today, however, the species still thrives – particularly in the Kalahari and Etosha national parks.

Left and far left: In the Kruger National Park, nyala occur in the remoter northern regions. Only the males carry horns, which are distinguished by smooth, ivory-like tips. Both sexes are coloured to merge with the dappled shadows of the forest glades they haunt.

A flight of Openbill storks.

Gregarious red lechwe occur in small herds of 15 to 20 individuals, but may congregate in their thousands in the Okavango Delta.

A white-fronted bee-eater, with a dragonfly it has just caught, perches for an instant before entering its nest, burrowed a metre or more into a riverbank.

In the Kruger National Park, an impala ram stands ever alert to signs of danger that could threaten his harem of ewes.

The Thomas Baines Nature Reserve in the Eastern Cape provides a habitat of mixed bush and Cape fynbos for this small group of eland.

A herd of red hartebeest in the Nossob River bed, Kalahari Gemsbok National Park. These gregarious antelope may occur in larger herds of 300, and in Botswana have been known to collect in their thousands at certain times of the year. Territorial hartebeest bulls vigorously defend their territories and harems (with their offspring) against trespassing males. These harem herds are relatively stable groupings, lasting for about three years. Males also form bachelor herds, although they have to make do with less favourable grazing, as the harems occupy the best areas.

The male nyala calf can be identified by the clearly discernible white chevron markings between the eyes. Mkuzi Game Reserve.

Created by the rising waters of the Zambezi, Starvation Island in the Kariba Dam – named during the famous wildlife rescue project of the early 1960s – is refuge to a herd of impala. Elephants and larger antelope are known to visit the island by swimming from the mainland, about 1 km distant.

Moremi Wildlife Reserve encompasses one of the finest wilderness regions of southern Africa, embracing almost every type of habitat from floodplains to savanna woodland, where large herds of blue wildebeest are still common.

Tsessebe in the Doorndraai Dam (far left) and Potgietersrus nature reserves (left). Tsessebe in southern Africa belong to the hartebeest genus, which is widely distributed in sub-Saharan Africa. A sturdy, fleet-footed antelope, it favours open plains with occasional bush and light woodland. Tsessebe generally form small herds of about three individuals, although they may congregate in larger numbers to graze. Territorial males regularly patrol their territories, but fighting between males is uncommon, even during the mating season. When herding a harem, the male exhibits a ritualised slow, high-stepping stride.

Sparring between adolescent waterbuck males is a prelude to more serious combats in later years when, as mature adults, territories will be gained and defended. Waterbuck have one of the strongest fighting instincts among ungulates.

Left: The red lechwe is a swamp-dwelling antelope which occurs only in the marshy flats of the Okavango Delta and eastern Caprivi. Like waterbuck, the male red lechwe is territorial, but is inclined to use bluff tactics for defence rather than physical engagement.

Top left: From a rocky promontory in the Kruger National Park, a bull kudu looks out over the dry bushveld. Kudu are also widespread in farming regions, especially in the central and northern parts of Namibia and the Transvaal and Zimbabwe lowveld. Their remarkable agility enables them to jump fences up to 2 m high.

A kudu and calf at the Msinga waterhole in the Mkuzi Game Reserve.

Baobab shadows fringe the shores of a salt pan in north-eastern Botswana.

The powerful bill of a marabou stork makes it an efficient scavenger. Like vultures, these birds rest for long periods in trees, but may also be seen stalking across the veld.

Top: Ensconced on its rocky outcrop, a klipspringer ewe in the Kruger National Park stays alert for danger. Should anything alarm her, she will 'freeze' in a typical pose on the tips of her hooves, barking when an intruder is spotted. If truly threatened, she will bound off to safety. The klipspringer's greyish-brown coat provides ideal camouflage in its rocky habitat, and its agility usually ensures a successful flight. Klipspringers live in small family parties.

The blue duiker is the smallest of southern Africa's antelope, weighing only about four kilograms. It is a forest-dwelling, nocturnal browser, resting up in the deepest parts of the forest during the day. Because of its shyness, little is known of the blue duiker's habits, but it is probably territorial, marking its area with secretions from facial glands.

Hamerkops about to mate in the Kruger National Park. These birds are fairly common throughout the sub-continent wherever there is open shallow water. They are generally solitary in habit.

Grey duikers (shown here at St Lucia and in the Lake Kyle region) are probably the most successful of all antelope, as they have adapted to a variety of habitats. In the wild they are diurnal, but have become nocturnal in places where they have been hunted or disturbed. Although they are predominantly browsers, grey duikers also catch and eat young guinea fowl and francolins.

Within view of the Langeberg Mountains, a herd of bontebok grazes peacefully in the National Park that was created to conserve the species in its natural habitat.

Nyala cows at a waterhole in Mkuzi Game Reserve. This antelope may be confused with a kudu cow, but is smaller, and has a darker rust-coloured coat with more well-defined stripes running down its flank. It also lacks the kudu's shoulder hump.

The Red-headed weaver is a common but localised species of the dry savanna. It builds its nest in a tall tree near water.

Above: A bull eland with a nursery herd in the Rob Ferreira Game Reserve.

Top and top left: Gemsbok are gregarious animals, occurring in mixed herds or nursery groups, although bulls may sometimes be solitary. Males form the largest territories of all the ungulates, and are more tolerant of trespassers than any other territorial species. How long they retain these territories is unknown. After extensive periods of isolation, a bull may join a mixed herd and enter strange territories with impunity, becoming subordinate to the territorial male.

Far left: A gemsbok herd on the dry, salty flats of the Etosha pan.

Inbreeding among the scattered groups of gemsbok in the Skeleton Coast Park has resulted in an unusual fawn-brown body coloration in place of the more familiar grey and black.

With conspicuous white tails flashing, a herd of black wildebeest gallops across the veld in the Karoo National Park. This species, endemic to the grasslands of South Africa, was brought close to extinction through hunting and habitat destruction caused by agricultural development. Recent conservation endeavours have, however, succeeded in creating viable herds in a number of reserves.

Left: By the shores of Kariba, a saddlebill stork takes advantage of the misfortune of a fish eagle that has just dropped its catch.

On the Chobe floodplains, a herd of waterbuck exhibit the characteristic white ring markings on the rump. These water-loving antelope form herds of 6-12 animals, but may also associate in larger groups of up to thirty. Waterbuck breed throughout the year. The female takes great care to hide her calf in tall grass, visiting only to feed and clean it.

Opposite: As in many antelope species, only male waterbuck carry the magnificent lyre-shaped horns, which are heavily ringed but smooth towards the tip. At age five or six, males reach a physical stature which enables them to hold territories. These territories are not relinquished during the year, and trespassers are warned off by a series of intimidatory displays in which the territorial male stands erect with head up, facing or side on to the intruder.

At Gorongoza, a pair of oribi test the wind, muscles taut in preparation for flight. These small antelope are found singly or in pairs in open habitats. The male has short, slightly curved horns.

Impala lambs amid rich savanna pastures.

When a predator threatens, both sexes of grey rhebok snort to warn others in the family group, maintaining the noise for up to 45 minutes if the source of the disturbance is not identified. These antelope also communicate by moaning and hissing. Thomas Baines Nature Reserve.

In late afternoon and at night, mountain reedbuck descend from the hillsides of the eastern Cape to feed on the valley grasses.

The Blue wildebeest is sometimes referred to as the 'brindled gnu' – a name derived from the dark bars spread across its body.

A herd of blue wildebeest disturbed while drinking at the Savuti channel in the Chobe National Park. Any unfamiliar sound or sudden movement at a waterhole is likely to put a herd of wildebeest to flight. However, they will often re-group a short distance away, and return warily to the same spot.

The Cape grysbok is a small, shy antelope which favours thick scrub bush, coastal forest or dry succulent veld. Active at night, it rests during the day, and is occasionally seen in the shelter of a shady bush.

Below left and far left: Red duiker in the St Lucia estuarine forest. Restricted to the eastern parts of the continent, from the Sudan southwards to Natal, red duiker prefer forests, dense coastal brush or thickly wooded ravines. Unfortunately, these habitats are being destroyed by agricultural development, and the red duiker's future is precarious. It is a small, solitary browser, with a loud penetrating call.

Sunset over the Zambezi.

A herd of blesbok gallops across the grasslands of the Bloemhof Dam Nature Reserve. These antelope are related to bontebok, sharing many of their habits, and from a distance they even look alike. The two species are thought to have descended from common ancestors. When the Karoo climate changed and became drier, the group split, leaving the bontebok in the southern Cape and the blesbok in the north. These grazers are gregarious, but tend to associate in small bachelor or breeding herds. The males are strongly territorial, defending their territories aggressively during the mating season. The antelope is named from the Afrikaans word *bles*, which describes the white blaze on its face.

Eland in the Mlilwane Wildlife Sanctuary, with cattle egrets in the foreground. The eland is Africa's largest antelope. Because it is resistant to drought, parasites and many diseases that affect cattle, much thought is today being given to commercial farming of captive herds.

Far left: Kudu bulls are noted for their magnificent horns. Although these antelope live in thick bush, their horns seldom restrict their movements: in flight, a bull lifts its head, laying the horns along its back, and plunges through thickets with apparent ease.

Left: A nesting cattle egret with its chick.

Above: The spiralling horns of the sable antelope, razor sharp at the tips, are deadly weapons when used against predators or in mating season clashes. Typically, the male kneels and scythes at his opponent.

Far left: At sunrise, impala gather on the sandy banks of the Shingwedzi River in the Kruger National Park.

Left: The hilly thickets and dry savanna woodland of the Zululand game reserves provide an ideal habitat for the stately nyala. This adolescent bull in the Hluhluwe reserve displays the characteristic white chevrons between its eyes.

Overleaf: In the Kalahari National Park, a herd of blue wildebeest congregates at a water hole in the early hours of morning. Wildebeest are totally dependent on the availability of water and never stray far from pans and rivers.

A sable calf displays its conspicuous chestnut coat, which will later darken to the glossy black of adulthood.

Above: A malachite sunbird feeds on a pin-cushion protea.

Top left: Sable antelope prefer dry woodland regions in reasonable proximity to water. Both sexes have spiralling horns, but those of the male are longer and more curved.

Top right: A steenbok eats the green shoots that quickly appear after veld fires in the Kruger National Park. These small, graceful antelope are a common prey for many predators.

Left: Sable become sexually active at 2½ years. Gestation lasts eight months and, in Zimbabwe, calves are born in the late summer. Victoria Falls National Park.

A young kudu bull among the mopani trees in the Hwange National Park. Kudu bulls are often found on their own, or they may join other bulls to form a bachelor herd. During the breeding season, a solitary bull may try to overthrow a territorial rival in a fight which can be fierce and even fatal – especially if the bulls lock their horns and are unable to free themselves.

Top & Top left: In the late 1800s, red hartebeest, like the bontebok and Cape mountain zebra, were in danger of extinction. Fortunately, remnant herds were preserved by concerned farmers in many areas. Today, red hartebeest are plentiful on the dry open grasslands of the Kalahari Gemsbok National Park (2). They also occur in Namibia and Botswana, and a herd has been established in the Suikerbosrand Nature Reserve, shown here (3).

Left: A roan antelope with its calf in the Percy Fyfe Nature Reserve. Both males and females of this species carry horns (the latter's are slightly smaller). Although gregarious, they are not found in large herds. Immature bulls form bachelor herds, while old bulls tend to be solitary. Most herds are nursery groups of females and their young, led by a master bull.

A small herd of Sable antelope gallops past a towering termite mound in the Kruger's bushveld.

Blue cranes are usually found in pairs, occasionally in large flocks. Their range covers the Karoo and open grasslands – mostly within South Africa.

Overleaf: Bontebok graze peacefully in the national park specially created for them near Swellendam in 1960. Whereas at the turn of the century their numbers in the wild were an estimated 300, by 1930 they had declined to a mere 20 individuals. Last-minute conservation efforts were undertaken with remarkable success, and today the species numbers over 360 in this park alone. Surplus animals have been translocated to other Cape reserves.

Impala are more susceptible to drought than most antelope. They are totally dependent on water and the availability of browse. Moreover, being herd animals, their competition for existing food is intense. The physical condition and number of impala are regarded as indicators of prevailing veld conditions.

Predators and Scavengers

Twenty-seven species of mammalian predators, from lions to serval cats, from mongooses to shrews, occupy the ecological niche at the top of the food chain. They are the hunters – the meat-eaters. All are perfectly evolved hunting and killing machines.

A carnivore's head says much about its way of life. The eyes are well forward, unlike those of a herbivore, which has eyes on either side of its head. Binocular vision is essential to a predator, as it must not only be able to see and identify its prey, but also catch it on the move. For that it needs to judge distance, speed and the right moment to strike. A one-eyed lion or cheetah has little chance of success in a solo hunt.

When a carnivore attacks, its claws secure the victim while its teeth are used to kill, then eat the prey. The big predators have developed varying techniques: a lioness may either suffocate her prey by seizing its nose or, like the cheetah, simply strangle it with a prolonged bite at the throat. Wild dogs and hyenas dismember and kill simultaneously. Knife-edged molars slice up the meat and, in the case of the hyena, grind up most of the bones.

The carnivore's food is more concentrated than a herbivore's plant fare. A kilogram of wildebeest food (grass) contains only a few grams of protein; the wildebeest flesh consumed by the lion is almost pure protein. Consequently, herbivores spend a great deal of time feeding, while predators eat in spaced bouts. They are sustained in the interim by their concentrated diets. Wildebeest spend 18 out of 24 hours finding food, while lions are likely to spend the same time sleeping.

However, this does not mean the lion is an habitual sluggard: hunting expends considerable energy, and there is no guarantee of success. Nor is there co-operation among the carnivorous species. Herbivores might associate to increase their chances of detecting danger, but recent research has shown considerable hostility among predators of different species. In addition, the hunters are dependent on the availability of their prey: when game is scarce, their survival is precarious.

It is axiomatic that there are many predator species because of the variety of herbivores on which they prey. Similarly, there are large predators because there are large herbivores. Buffalo, zebra, wildebeest, impala and springbok make it possible for lion, leopard, cheetah and hyena to exist.

Under natural conditions – in areas vaster than any parks in southern Africa – there is a delicate balance between habitat, prey species and carnivores. Disturb one, and the repercussions become readily apparent. Because predators are at the top of the food chain, it may be assumed that they control the population sizes of the other species. In fact, the reverse is true. In a stable ecosystem there are always fewer carnivores than grazers and browsers. When prey becomes scarce, some meat eaters must inevitably starve.

Predators are selective killers, hunting the weak, injured or ailing, thereby allowing the fittest to survive. Sometimes they prefer a specific species. For example, impala – being very numerous – are taken frequently. Consequently, they form a 'prey buffer' for less abundant species.

In the wild, death is part of the natural order. A herd of zebra stampeded by attacking lionesses will stop and continue grazing once the victim has been isolated and killed. They know instinctively when the moment of danger has passed, and when they are no longer at risk. Nature does not recognise sentiment.

When one sees a lion in its natural habitat, it is understandable why this animal has always aroused such respect in humans. Certainly no other animal features as often in folklore, nor commands such respect – or such fear. Perhaps it is the sense of menace, of power, of intrinsic violence, that attracts Man. Perhaps the lion plucks a chord of primeval dread.

The lion hunts mostly at night, but – being an opportunistic killer – will also take advantage of an easy victim during the day. On the plains, where there is little cover, it is obviously in its interests to attack in the dark. As the sun sets, activity within the pride increases. The lions yawn and stretch, and the cubs play with more vigour. When the adults move off to hunt, the lionesses generally lead, for they are responsible for bringing down most of the prey. The females usually hunt as a team to improve their chances of success.

Typically, two lionesses flush the prey, driving it towards other lions stalking through the grass. In most cases a lioness catches the victim by knocking it over with the swipe of a paw, or by pulling it down. Small victims, such as hares or foals, are killed with a bite, while impala, wildebeest or zebra are frequently killed by crushing the mouth and nose until the prey dies of suffocation. Alternatively, the lion might bite down on the windpipe until the animal strangles. Wildlife researchers estimate that probably only two in every five attacks is successful.

For the tenacious lions of the Kalahari Gemsbok National Park, prey as large as a wildebeest would be a considerable luxury. Because of the harshness of their semi-desert terrain, they may prowl for many hours in search of a prize that may be a steenbok at best; at worst a rodent. Compared with lions in other parks, where small mammals constitute no more than one per cent of their prey, 50 per cent of the Kalahari lion's diet is made up of small mammals.

Should a Kalahari lion succeed in bringing down a gemsbok, he is most likely to kill it by landing on the antelope's back, digging his teeth into the haunch and jerking backwards, snapping the neck.

Although lions will drink once a day when water is available, they can go for up to a month without water. However, some Kalahari lions have never been known to drink.

Cheetah are a plains species, and their way of hunting is ideally suited to the savanna grasslands. With their excellent eyesight, they can identify prey from long distances, and even during the rest period (in the middle of the day) will sit up to survey their surroundings as many as ten times an hour. Cheetah are mostly diurnal hunters, favouring the early morning hours.

With their small heads, deep chests and long legs, cheetah are the least typical of the cat family. Their claws are only slightly retractable, functioning like an athlete's spikes to aid acceleration. Longitudinal ridges on the pads are another helpful adaptation – they prevent skidding at high speeds. Cheetah are built for speed: the flexible backbone and moveable shoulder blades allow the back legs to be brought far forward, giving the animals a phenomenally long stride.

When a cheetah sprints, there are two moments during its running movement when all four feet are off the ground. This happens when the legs are stretched out, and again when they are bunched. Each stride covers about seven metres. The acceleration rate is impressive: a cheetah can reach a speed of 72 km/h from standstill within two seconds. Within a few more seconds it can be running at a speed of 112 km/h – earning it the reputation of being the fastest mammal. The cheetah uses its long tail for stability at high speeds, swinging it much like a rudder to change direction. It follows that this speedster needs as clear a run as possible, with no obstacles such as trees or rocks to impede its attack.

The hunt inevitably ends in failure if the prey zig-zags, or runs into thick bush. The cheetah is careful not to start the chase too soon, or it will quickly exhaust its stamina. Unless it catches its prey within 400 metres, it may well give up. It could be 30 minutes before it regains its breath.

When a cheetah sees prey, it begins to stalk rapidly with raised head. If the prey becomes nervous, the cheetah continues to stalk with its body pressed against the ground. The approach is from behind, and continues until the cheetah is within 50 or 60 metres. The sprint is explosive, the cheetah either knocking over its victim or jumping on its back. Death is by strangulation or a bite on the jugular vein.

With its small head, weak jaws and teeth, a cheetah is restricted to hunting medium-sized and small antelope such as springbok, impala, reedbuck, duiker and the young of kudu, wildebeest and waterbuck. If game is scarce, it resorts to prey such as hares, porcupines or guinea fowl. Unlike the other big cats, the cheetah seldom scavenges.

Spotted hyenas are perhaps the most maligned predators of the plains. Legend has them filling the night with a malicious cackle, or circling a kill – treacherous, humpbacked, heavy-headed – in wait for the leftovers while lions feed. Certainly their 'whoop' is as famous as a lion's roar, and hyenas do seem to 'giggle' and 'laugh'. The 'giggle' is in fact a sign of fear, and they tend to 'laugh' with excitement at a kill.

Hyenas are related to cats, although in many ways they look and act like dogs. They have rounded, cat-like ears, but their claws are short and blunt and cannot be retracted. Because the female's external genitals are similar in appearance and often larger than those of the male, they have been wrongly labelled as hermaphrodites – a further smear to an already unfair reputation that identifies hyenas as cowardly, merciless ruffians of the night.

As scavengers, hyenas perform a valuable function. The veld would soon be a disease-ridden wasteland if it was littered with decaying meat and bones. Fortunately, hyenas – thanks to their massive heads and powerful jaws and teeth – can crush all but the biggest elephant bones. To break down bone splinters, hyenas have a remarkable digestive system. Once the bones are splintered, chewed and swallowed, the digestive system extracts all the organic matter, allowing only inorganic matter to be excreted. Consequently, their droppings are a conspicuous white colour – almost pure calcium, mixed with hair.

Recent research has shown that the spotted hyena is not merely a scavenger, but also a bold and resourceful hunter. Aided by its finely tuned senses of hearing and smell, and by good night vision, it seldom allows its prey to escape.

Hyenas usually hunt by night, preying on wildebeest, zebra, young antelope and even elephant calves. When hyenas attack as a pack, they may select a victim by running into a herd of wildebeest, then standing by to watch as the animals run off. A victim is then chosen and chased out of the herd. Sometimes the hunt may extend for only a few hundred metres; on other occasions it could range for 3-4 kilometres. Hyenas are strong runners, and can reach speeds of 40 km/h – which makes them formidable hunters. To bring down a fleeing antelope, the hyenas dart in, biting at its legs and even tearing strips of meat from its flanks. Eventually, in a state of shock and probably immune to pain, the victim is downed. Although it is not killed immediately, the animal dies very quickly.

The other species of hyena found in southern Africa, the brown hyena, differs in many respects from its relative. Although they are found in savanna regions, they also occur in the western deserts. Brown hyenas are the smaller species, with long, shaggy, brown coats, heavy manes and bushy tails. Where they occur in coastal areas, they hunt and scavenge on the beach, feeding on anything from stranded whales or seals to birds, dead fish, squid and even mussels. This habit has earned them the Afrikaans name of 'strandwolf' – beach wolf. They also hunt or scavenge at night. Brown hyenas are not as aggressive as spotted hyenas. In the Kruger National Park and in Botswana, they are reported to have attacked and killed a wide range of species, including such large animals as kudu. Normally, however, they hunt small antelope, rodents, birds, reptiles and insects, and they may also eat fruit and eggs.

Wild dogs co-operate in much the same way as a clan of hyenas. These creatures, little bigger than jackals, hunt with ruthless efficiency. Their kill rate is impressive as they do not employ the hit-and-miss tactics of lions and cheetahs.

The size of a pack varies from a few individuals to as many as forty. The pack usually hunts after sunrise or at dusk. First they confuse, distract and harass their prey until an animal is separated from the relative security of the herd. Once on its own, the victim has no respite. It is run down in a series of flanking movements which enables 'relays' of wild dogs to take over the chase as the lead dogs tire.

Their endurance is legendary; their silent determination terrifying. As their prey flags, they rush in, ripping at its sides until it falls. Occasionally their victim will be alive for several minutes while they feed. However, shock usually makes the animal insensitive to pain. In Hwange, a pack of about 40 dogs was seen to reduce two reedbuck ewes to a pile of bones in less than ten minutes.

Such hunting co-operation increases the amount of food available to wild dogs: they will eat anything from hares to zebra. This is the essence of their social hunting and is crucial to their survival. A hyena might be able to live as a solitary scavenger, but the frailer wild dogs are bound to their comrades – without them they will starve or be killed. Yet despite their social nature, wild dogs are not above cannibalism if one of their number is wounded.

Unlike the chasing tactics employed by the hunters of the plains, leopards rely on cunning and stealth. Once again, there is an analogy between the way they hunt and the type of habitat they favour. While leopards are found on savannas, they are also at home in forests and on mountains. Because of their varied diet and secretive nature, these creatures have adapted to many different habitats, and may even live close to cities. There are leopards in the Magaliesberg mountains, less than 100 km from Johannesburg, and in the Boland mountains not far from Cape Town.

The leopard is solitary, and hunts mostly at night. Its tactic is to stalk its prey, using whatever cover is available. Among the shadows of veld or forest its spotted coat provides excellent camouflage. Once within striking distance, it dashes the last stretch to pounce on the victim. A baboon or impala is killed with a bite on the back of the neck, which cuts the spinal cord and causes instant death. Larger animals such as wildebeest and hartebeest are probably strangled. Sometimes a leopard may ambush its prey from a tree. Stretched out along a branch, it remains well hidden and waits for an animal to walk beneath – then drops on to the unsuspecting victim's back and kills it. Like domestic cats, leopards can retract their claws when they are not needed. The strong, curved claws are bared when climbing, fighting or killing.

In areas where game is plentiful, leopards eat small or medium-sized ungulates such as impala, reedbuck, duiker, steenbok and warthog. They have also been known to take kudu, hartebeest and sable. A study in Zimbabwe revealed that their diet included dassies, rats, mice, hares, klipspringers, birds, baboons, snakes, lizards, insects and even scorpions. On an island at Lake Kariba, a leopard was seen catching and eating bream. In the Kruger National Park they are recorded as eating 31 species of mammal, both large and small.

Having eaten, a leopard drags the carcass into the high branches of a nearby tree, where it is out of reach of other predators or scavengers. Such is the leopard's strength that it can lift weights far higher than its own. A leopard returns repeatedly to the kill, even when the flesh is putrid. In fact, these cats will scavenge should they come across carrion.

Every predator has an enemy which might kill it for gratuitious reasons. Lions have to be wary of packs of hunting dogs or hyenas, and the young of all are at risk from other carnivores. Unlike the herbivores, predator species do not intermingle. To avoid competition and conflict, the hunters have different habits: they are nocturnal or diurnal, solitary or social. But as several predators favour the same prey species, clashes are inevitable. Although antelope may seem abundant, to the meat eaters there are only as many as they can catch. Consequently, it is not surprising that the most noble of predators will stoop to plunder when the opportunity presents itself. For example, it is possible that hyenas skulking in the vicinity of a feeding pride of lions could be the rightful 'diners' – driven off by one of the most opportunistic feeders on the veld.

Leopards can secrete their prey in a tree and feed at leisure, but cheetahs seldom have time for a leisurely meal. After the energy-sapping sprint, they eat quickly, glancing around nervously to ensure they are not being approached by lions, hyenas or wild dogs – all of which have no trouble in chasing cheetahs from their kill. To make the most of their meal, cheetahs eat the protein-rich muscles of the hind legs first. Other large predators invariably start with the soft belly, progressing to the rest in their own time.

Being solitary hunters, cheetahs do not have the advantages of the pack when it comes to defending their prey. Writing in the magazine *African Wildlife*, a tourist to Hwange described how six hyenas tried to steal a slain wildebeest from a pack of hunting dogs, but met with no success. Each time one approached too closely, a group of dogs would drive it away.

Had the hyenas attacked as a pack, they would probably have fared better. Hunter Gordon Cundill recalls such an instance which was all the more remarkable for the cunning displayed by the wild dogs. Beaten off, they waited while the hyenas divided up the kill, each dragging away a share. Once the hyenas had separated, the wild dogs attacked as a pack, methodically harassing hyena after hyena until they had recaptured all the pieces of their kill.

Of all the carnivores, lions are the most intolerant in their dealings with other predators. They will stalk and chase them, sometimes killing without provocation, and without warning. The less agile brown hyena and aardwolf often fall foul of lions, and spotted hyenas, cheetahs and black-backed jackals are occasional victims. Only recently have wildlife researchers realised the extent of killings among animals at the top of the food chain.

When lions kill other predators, they are not intent on a meal. At most, they might partly consume the victim, but usually they worry at it briefly, then stalk off. Interestingly, they attack other carnivores with the bared teeth and snarls usually reserved for dealings with their own species. Only very infrequently do lions kill other lions. They are only too aware of their opponent's capacity to inflict mortal injury.

Most members of the cat family are solitary in habit, but lions are social animals, living in prides made up of several adult males, females, and cubs of various ages. Nomadic individuals are usually outcasts, or injured. Adults may also form 'bachelor' groups.

The size of a pride depends largely on the availability of prey animals. In areas where prey is abundant, prides may number as many as 30 members. The lionesses keep the pride together, as adult females are usually closely related – being either sisters from the same litter or daughters and granddaughters. Most of the leadership within a pride comes from the lionesses, but the males still dominate by virtue of their size and strength.

Lions are not strongly attached to the pride, and remain only as long as it suits them, or until they are driven off. When a male is ready to wander, he is unlikely to resist another lion's challenge to take over the pride. Should the dominant male not give way, the fight may be vicious and could possibly end in death for one of the combatants. The male's wandering habit is important, even vital, as it allows the introduction of new blood into prides where all the females are related, thereby preventing inbreeding that could weaken a population.

The dominant male establishes a territory for the pride which he marks by spraying urine on bushes. A territory varies in size according to the number of prey animals, and on the availability of water in the region. Both males and females protect the territory by fighting off intruders of their own sex.

Male invaders are most likely when females are on heat. If a fight occurs, it may be long and fierce. Inevitably, both combatants are left torn and bleeding, and sometimes both may die from their wounds. Should a strange lion manage to defeat a territorial lion, he takes over the region and is automatically accepted by the females.

A lioness about to give birth leaves the pride to find a secluded lair among rocks or in thick vegetation. There the cubs are born and kept hidden. At first, their coats are spotted. This camouflage is important, as the litter remains in the lair for six weeks before rejoining the pride. During this time the lioness has to hunt and care for her young; the males share none of this responsibility.

At the age of about six weeks, the cubs are taken to a kill and allowed to eat soft meat such as the liver and intestines. They are now also old enough to learn to hunt, which they do by imitating their mother. After about six months, the mother starts giving more attention to the hunt than to her cubs, and it is up to them to follow as best they can. Even at the age of 16 months they are dependent on their mother, but after two years independence is virtually forced upon them, for by now their mother will be pregnant again. By that time they are capable of hunting for themselves.

If food is plentiful, lions will gorge themselves on a carcass and move off. Otherwise, they drag it into the shade of a tree, where it is guarded until the next meal. Although the lionesses make the kill, it is immediately appropriated by the males. It is only when they are finished that the females can have their share. The cubs and younger members of the pride have to wait their turn. Should a cub, overcome by hunger, attempt to steal a mouthful, it may be dealt a powerful blow by an adult – sometimes with fatal results.

Whether a cub gets a chance to eat depends on the size of the pride and the size of the prey. Lions, unlike leopards – which may kill for reasons other than hunger – kill only sufficient to satisfy their needs. Consequently, many cubs die of starvation. In addition, lions on the prowl traverse large distances, and the pace soon takes its toll of undernourished cubs.

Ironically, the much-maligned hyenas show more consideration for their offspring. Female hyenas give birth to their young in a den, which could be an old aardvark burrow. The cubs are suckled for the first 18 months, and by the age of two years may be eating meat from kills made near the den. If the kill is too distant, the female regurgitates meat for her cubs.

There is no pecking order among hyenas at a kill. The prey is literally torn to pieces, the hyenas fighting for a share, and scattering in all directions. Each hyena finishes its portion in solitude.

Wild dogs, with their penchant for co-operation, display this quality in all facets of their lives. They live in packs which may number as many as 30 individuals. Although wild dogs are great wanderers, when a female gives birth, the pack establishes a base near the den. From here the adults conduct their hunting expeditions until the pups are strong enough to stand the pace.

Litters consist of two to six pups, few of which survive: disease and marauding lions and leopards take a heavy toll. For the first three months, the pups suckle. They venture out of the den when aged only 2 weeks, spending much of the time playing together.

When the pack goes hunting, the mother stays with the babies (she is sometimes replaced by another adult). On returning, the pack regurgitates food for the adult. Once the pups are weaned, they too are fed on regurgitated meat.

The pack spends most of the day lying near the den. In the late afternoon, they get ready for the hunt with excited twitters and squeals, and much nosing and licking of each other's lips, tails wagging furiously. This seems to put them in the right hunting mood. There is usually one adult leading the pack each time, deciding where to go and which animals to chase.

At about the age of 2½ months, the cubs leave the den to hunt with the adults. They are frequently left behind during the chase, but are never abandoned. Should the cubs not catch up an adult returns to fetch them. They are allowed to feed first, while the adults stand around them to keep other predators away.

A wild dog kill is remarkable for the control exhibited by the pack. There is no frenzy, no snapping or fighting. The dogs eat in their hierarchical order, each waiting its turn – apparently secure in the knowledge that if there is not enough to go around, the pack will make another kill.

Cheetahs also take considerable trouble in rearing their offspring. Being solitary hunters, they face possibly greater hardship, and in the first few months half the litter may die from disease or lack of food, or be killed by predators.

A cheetah's litter consists of three or four cubs which the female raises alone. She hides them in a lair and frequently moves them to new lairs to prevent their smell becoming too strong in one place, thereby attracting predators. By the age of eight weeks, the cubs are following their mother constantly. This is a difficult time for the female, as the cubs may spoil her chances of catching prey. Once she has made a kill, she summons her cubs with a high chirruping sound. The cubs eat while the female regains her breath and watches for other predators.

Cheetah cubs show an interest in hunting from an early age, and like other cats, have to be taught to stalk, pull down and kill their prey. Training begins when they are about eight months old. The mother will catch a small or young buck and release it near her cubs, leaving it to them to catch and kill it. This performance may be repeated many times with the same victim.

Young cheetah can stalk and kill successfully once they are about 18 months old. They are now ready for independence, and will leave the family group abruptly – unlike lion cubs, which learn independence more slowly.

Apart from the big cats, hyenas and hunting dogs, there are many other smaller predators. However, they tend to be solitary and nocturnal, and are therefore seldom seen. Their prey varies from small antelope to birds, mice, hares, lizards, poisonous snakes and insects.

Most savage and powerful among the smaller predators is the caracal. This cat has many leopard-like traits: it can also be extremely dangerous when wounded, and sometimes drags its food into trees.

African wild cats, which are a little smaller than caracals, were probably the original ancestors of domestic cats. They were tamed by the ancient Egyptians, and were common house pets in Mesopotamia and Palestine 6 000 years ago. These cats sometimes interbreed with domestic cats, but the offspring always have the black striping of the wild cat and inherit its solitary, independent habits.

A correspondent to *African Wildlife* records how he watched a wounded wildcat in the Kruger National Park struggling back to her lair and litter. Making painful progress, she reached them, lay down and washed her kittens. The ritual complete, she killed each one with a sharp bite at the back of the neck. Then she died.

Genets, which are more closely related to mongooses than cats, were also tamed by the ancient Egyptians (they are far better rodent catchers than cats). Genets are solitary and nocturnal, and agile tree climbers. In a way, they are like cats: the young have a distinct purr, while threatened adults hiss and arch their backs like domestic cats.

Unfortunately, genets are regarded as a nuisance by farmers, as they supplement their diet of birds' eggs, mice, lizards and grasshoppers with poultry if the opportunity arises. The conflict between small predators and farmers is inevitable, yet difficult to resolve.

Black-backed jackals are probably the farmers' worst enemies, as they kill sheep, goats, lambs and poultry. They even raid vineyards, orchards, sugar cane and pineapple plantations. Because they nibble at the fruit, they spoil far more than they can actually eat, so farmers tend to shoot them out of hand, or lay poisoned traps.

Black-backed jackals have also been accused of spreading rabies, and many are killed in the annual campaigns against the disease. Because poisoned baits are used, many other predators and scavengers are also killed – a source of much contention among conservationists. They maintain there is no scientific evidence that black-backed jackals do spread the disease.

These jackals are members of the dog family, and closely related to wolves. They are cunning, intelligent animals, and manage to thrive even though they are hunted relentlessly by farmers. In the wild, their main enemies are leopards, cheetahs and occasionally lions. Black-backed jackals seem to be both nocturnal and diurnal, and they may live singly or in pairs.

Jackals give birth in an old aardvark burrow, a hollowed-out termite mound, or in a rocky crevice. The litter of 4-6 blind, helpless pups is suckled and fed on regurgitated meat until the age of about four months, after which the parents bring in small animals. Because the pups remain at the den until they are fully adult (at two years of age), they often help to rear the cubs of the next litter.

Black-backed jackals are both scavengers and predators. At a lion kill, they hover in the background, waiting for the large carnivores to finish eating. Even then they have to snatch what they can in competition with hyenas and vultures.

While no predators have become extinct in southern Africa, several are severely threatened. Wild dogs are scarce, having suffered badly at the hands of farmers. They have been exterminated in most of South Africa, and now survive only in the game reserves of the eastern Lowveld. In the Kruger National Park, they number no more than five hundred.

Cheetah, too, faced a similar problem: until recently, they were unprotected in the Transvaal. Ironically, while there are relatively few cheetah in the wild, they breed prodigiously in captivity. Successful re-introduction to areas where they were once extinct now means their survival is no longer at risk.

In the long term, the survival of all predators depends on Man's commitment to conserving the natural environment. As long as there are sanctuaries, wildlife will thrive. But in a world of burgeoning human numbers and political strife, keeping them inviolate might prove at best controversial, at worst impossible.

Previous Page: In Nwanedi National Park, a lioness and her cubs take shelter from the hot sun.

Two young cheetahs assess their chances of making a kill. Cubs become independent when about one year old, and may suffer lean times until they acquire hunting skills.

Left: Males of a litter sometimes remain together into adulthood, forming small bachelor parties which may be joined by other males. Usually, however, cheetahs are solitary creatures. Sexually mature females live alone, except when they have cubs.

Far left: Two cheetahs in the Kruger National Park set off on the hunt. These predators hunt mostly in the early dawn or late afternoon, as they need good visibility, and must avoid competition with predominantly nocturnal killers such as lions, hyenas or leopards.

Above: The honey badger (or ratel) is one of the most unmistakable species of the veld, with its greyish-brown saddle and jet-black underparts. The claws on the forefeet are long and powerful, like curved knives. These claws are put to effective use when tearing beehives apart, or digging for prey. Although bee larvae and honey gave the honey badger its name, scorpions form the bulk of its diet, followed by mice, spiders, insects, small birds and snakes. Its thick skin and coarse coat protect the honey badger when it raids a beehive.

Far right: Showing its formidable claws and teeth, a honey badger pauses in its search for insects. This animal is extremely courageous, and has been known to attack wildebeest and waterbuck for no obvious reason. It will tackle a snake as large as a 3 m python.

Right: The Wood owl, distinguished by its lack of 'ears', is a localised resident of dense forest and exotic plantations.

Little attention was given to the white lion phenomenon, despite the sighting of two cubs in 1959 (near Tshokwane in the Kruger National Park), until a white male was seen in this area in 1976. Later that year, three white cubs were born in the Timbavati game reserve on the Kruger's western boundary. Since 1979, at least nine other white lions have been born in this region. Despite their conspicuous colouring, they are accepted by the pride. The white lion phenomenon is due to a recessive gene in local prides which produces individuals lacking the melanin pigment. All litters featuring white cubs have also had normally coloured individuals, as shown here (left and far left) in Timbavati. The white coats tend to darken with age.

Melanism is not common among wild animals, but does occur from time to time, producing black creatures such as the genet shown here. Two species of genet occur in southern Africa: the small-spotted genet, which favours arid areas, and the large-spotted genet, usually found in riverine forest. Genets are solitary, nocturnal carnivores.

Below: In many respects the serval, with its long legs, slender body and yellow-barred and spotted fur, looks like a small cheetah. This solitary, nocturnal predator has the good hearing, sight, speed and agility it needs to hunt successfully for birds, rats, hares, lizards and insects. Hwange National Park.

Right: Marsh owls frequent the long grass in the marshy Okavango swamps.

Far Right: An African wild cat in the Moremi Game Reserve. More than 8 000 years ago, tamed wildcats were common house pets in Egypt, Mesopotamia and Palestine. Wild cats often interbreed with domestic cats; the kittens invariably inherit their father's black stripes and solitary, independent ways.

Top: Black-footed cats occur in dry, open country where they prey at night on insects, reptiles, small birds and mammals. These cats are smaller and less common than African wild cats, but have a reputation for being more aggressive.

Wild dogs are gregarious, forming packs of 10-15 individuals. Sometimes, however, they may number as many as fifty. Of all the carnivores, these creatures are probably the most successful hunters, renowned for their endurance and co-operation. The blotched markings on their coats are a useful camouflage as they stalk through the veld.

A hooded vulture scavenges the remains of a buffalo carcase.

Left: Wild dog pups in the Kruger National Park. Although the average litter size is 4-8 pups, the mortality rate is high because of disease and predation by lions, leopards and hyenas. To afford maximum protection, the litter is born in a den – usually a hole in the ground. By the time the pups are weaned (at about three months), they will be hunting with the pack.

A young male lion at its kill in the Timbavati Nature Reserve. Lions kill and eat a variety of animals, but prefer medium-sized grazers such as zebra and wildebeest. Being at the top of the food chain, they appear to lead less demanding lives than herbivores, which must spend many hours foraging and be constantly alert for predators. Yet the lions also work hard to survive: their average kill rate is only two in five.

Following page: A group of lions slake their thirst at night at a waterhole in the Mala Mala Game Reserve.

Left and above: Leopards occur in a wide range of habitats, and although they prefer rocky hills, mountains and forests, they may also be found on the fringes of semi-desert regions. Because they get most of their liquid requirements from their prey, leopards are independent of water – though they will drink regularly when it is available. They are better adapted to forests than open savanna, where they tend to keep to thick bush or rocky terrain. Males establish and defend their territories against other males, resorting to combat if the transgressor ignores the warning grunts and growls. Females are allowed to range through a male's territory. Although solitary creatures, they may be seen in pairs during the mating season or as a family of female and cubs. Young leopards become independent at about 22 months. However, 'affectionate reunions' between mother and offspring have been reported.

Above left: Barn owls are common throughout southern Africa and breed in a variety of localities – trees, caves, nests of hamerkops as well as the eaves and nooks of buildings. They are voracious nocturnal feeders on small rodents and hence of great benefit to farmers.

Far left: The Okavango Delta wilderness is one of the few remaining habitats of Pel's fishing owl. This large and distinctive raptor is mainly nocturnal, but can sometimes be observed singly or in pairs in the twilight hours.

Right: A black-backed jackal prowls a stony stretch of the Skeleton Coast. Although this carnivore is known mainly as a scavenger, it is also an effective hunter: in this region it preys on seal pups. The black-backed jackal also eats hares, rodents, birds, reptiles and insects. In lusher areas it will raid vineyards, orchards and sugar-cane and pineapple plantations.

Far Right: Black-backed jackals are found throughout southern Africa, although they favour the more arid areas.

Below right: The side-striped jackal is mainly nocturnal, although it may be seen in the early morning or late afternoon. During the day it hides among rocks or in abandoned aardvark holes. This species favours woodlands rather than the plains, where black-backed jackals occur.

Below: The spoor of brown hyena and black-backed jackal on a Skeleton Coast beach.

Lion cubs remain with their mother for some two years, by which time she is ready to mate again. They are weaned at about seven months, but at this stage are still unable to fend for themselves, and rely totally on the mother's hunting skills. Cub mortality is high, and usually only half the average litter of four will survive. Mala Mala Game Reserve.

Above & Far left: According to the fossil record, spotted hyenas are more closely related to mongooses and civets than to the dog family. Although they have round, cat-like ears, their claws are strong and blunt like a dog's, and cannot be retracted. They also behave and look a little like dogs. Set in a massive head, the hyena's powerful jaws are reputed to be the most powerful among the mammals, and can crush all but the largest of elephant bones. They are equipped with 34 teeth for grinding bones and slicing flesh, and their strong digestive juices can dissolve most matter. Hyena droppings are conspicuously white, consisting almost entirely of calcium.

Left: The Yellow-billed hornbill is a common species of dry savanna thornbush and mopani areas. It spends much of its time on the ground searching for insects, fruit and seeds, and eats small vertebrates such as lizards.

Cheetahs at the Loskop Cheetah Sanctuary. Despite its ferocity, the cheetah cannot match the strength of the other large carnivores, so this predator is restricted to hunting medium- and smaller-sized antelope such as springbok, impala, reedbuck and duiker. Cheetahs may also take young kudu, wildebeest and waterbuck, and will kill hares, porcupines and even guinea fowl if other prey is scarce. Unlike other big cats, they seldom scavenge, as they prefer freshly killed prey

With its small head, deep chest and long legs, the cheetah is built for speed – but it needs plenty of open space when hunting, as it is unable to change direction quickly. The cheetah has little stamina, and must make its kill within 400 metres, or give up. The cheetah's heavy tail acts as a 'rudder' during the sprint.

A lioness and her cubs at her buffalo kill in the Mala Mala Game Reserve. Usually, the belly is torn open and the soft stomach and intestines eaten before attention is given to the protein-rich parts of the kill. Lions have been known to draw the intestines through their incisors, thereby squeezing out the contents. Because of a rapid digestive process, they are able to take a second meal shortly after the first gorging.

Right & Bottom centre: Brown hyenas are commonly regarded as solitary carnivores, probably because they forage alone. But researchers in the Kalahari have found that they form clans of 10-12 individuals made up of a dominant male, a number of subordinate males and four or five adult females with young. Males maintain their social position by ritual fights such as muzzle-wrestling and neck-biting; although blood is drawn, injuries are seldom serious or fatal. While the ranges of brown hyenas and spotted hyenas overlap, the former occur mainly in the arid western areas. They are less aggressive than spotted hyenas and are unlikely – even as a pack – to kill ungulates as large as kudu or zebra.

Below & Bottom right: Spotted hyena cubs are totally dependent on the mother until they are weaned, when about one year old. The cubs start to follow the mother during a hunt when only a few months old, but do not actually join in the kill until fully grown.

Although predominantly nocturnal, the aardwolf can be seen in the early or late daylight hours as it searches for termites. It moves slowly when foraging, relying on its acute senses of hearing and smell to guide it to where termites are active. Using its front claws, the aardwolf first excavates termites from a mound, then laps them up with its broad tongue, which is covered with sticky saliva. This creature has formidable canine teeth and a jaw strong enough to inflict a severe bite. However, it will usually deter an aggressor by roaring loudly and erecting its mane. Although there have been reports of aardwolves at a carcase, the animal's incisors are weak, and would certainly not cut meat. It is more likely to favour maggots and carrion insects. The aardwolf occurs singly, in pairs or in small family parties. During the day it lies up in a disused aardvark burrow or the enlarged holes of other species, such as the spring hare. The Nxai Pan National Park is one of the best localities to view the aardwolf.

Previous Page: A lioness at Mala Mala Game Reserve with her three young cubs. Up to eight weeks after birth, the lioness remains apart from the pride, hiding her young in clumps of bush and tall grass to protect them from other predators, including male lions (even her own mate). The cubs are weaned at about seven months, but remain dependent on their mother for another 18 months.

Middle: A Spotted eagle owl chick.

Left: With its keen sight and acute hearing, the small-spotted genet is well suited to nocturnal hunting. When threatened, it defends itself by arching its back and erecting its spinal hair in a manner similar to that exhibited by domestic cats. It is a proficient climber, and will take to trees if hunted.

Bottom left & far left: The caracal's strong, sharp teeth are set in a powerful jaw which is well adapted to deliver a killing bite and slice up the meat. This carnivore preys on small and medium-sized mammals such as dassies, ground squirrels, steenbok, the young of larger antelope, and birds such as guinea-fowl. Because it is secretive and prodominantly nocturnal, little is known about the caracal's behaviour. It is a solitary creature, mating for short periods only. The female gives birth to a litter of 2-4 kittens in an abandoned ant bear hole, or in a hole in a tree.

A pair of mating lions in the Timbavati Game Reserve. Oestrus lasts from 4-16 days, and during that time the lioness remains in close association with a male. Copulation may be initiated by either member of the pair: the female may take the lead by crouching in front of her mate, or the male may give a ritual snarl, wrinkling his nose and exposing his canines (as shown here in far left). Lions are recorded as mating every 15 to 20 minutes over a period of several hours; they sleep or lie next to one another when not mating. Aggressive behaviour during copulation is a feature of many carnivorous species, but the lion's snarl at the end of a mating bout is ritualistic, as is the gentle neck bite (left).

Previous page: Lean and sleek, the fastest predator on the African plains, a 'king' cheetah circles warily at De Wildt. This individual shows the fused stripe-like markings which led some scientists to believe it a cheetah subspecies. However, it is now accepted that this is merely a colour variation which occasionally occurs among cheetah in Zimbabwe and South Africa. By selective breeding between individuals with the colour variation, it is possible that 'king' cheetahs will become more common. They cannot be awarded subspecies status, as they do not occupy separate geographic areas.

An African fish eagle adopts a characteristic tree-top lookout post over the Okavango River, where it preys on bream and catfish. Its high-pitched, ringing call is one of the most evocative sounds of wild Africa.

Play is an important aspect of a lion cub's growth. It romps incessantly, sometimes arousing the wrath of adults. Here a cub snarls from atop a termite mound in apparent challenge to its fellow cubs.

A lioness with her cubs beside the Makwa Pan, in the Hwange National Park. She looks up from her buffalo kill on hearing a disturbance in the nearby bush

Although mainly nocturnal, bat-eared foxes may be seen foraging in quiet areas during the morning or evening. These creatures have as many as 50 teeth (more than any other land mammal). However, the teeth are weak and suited only to a diet of insects, scorpions, snakes, lizards, small mammals, fruit, berries and soft roots.

At Nxai Pan National Park, a pair of bat-eared foxes lie up during the heat of the day.

Using its sensitive ears to detect grubs and beetles in the soil, a bat-eared fox in the Kalahari National Park pauses before digging out its prey. These large ears also serve as 'radiators' which help to maintain an even body temperature.

A bat-eared fox at Nxai Pan National Park searches about a termite mound for insects.

Late afternoon sunlight filters through the acacias at Mana Pools National Park.

In Timbavati Nature Reserve, one of the famous white lion cubs yawns at the end of its afternoon rest, a gesture duplicated by a black-maned male. The yawn is not so much an indication of fatigue as a sign (along with increasing restlessness) that the pride may soon move off on the hunt. Although the males – by virtue of their size and strength – dominate the pride, it is the lionesses which take the lead and do the actual hunting. White lions are an anomaly in the bush. Their conspicuous colouring makes them highly visible, lessening their chances of hunting successfully.

Small Game

Mongooses are undoubtedly the most ubiquitous carnivores of southern Africa. These small, sleek animals, with short legs and well-formed tails, are often seen standing like sentinels in the veld, watching for danger. There are some 75 species scattered from southern Europe to Asia, Madagascar and Africa.

Mongooses differ from cats in having a fifth toe on the hind foot, non-retractable claws, and more teeth (36-40). When threatened, they secrete an obnoxious fluid.

Most of the mongoose species are diurnal and gregarious, but the Cape mongoose occurs in pairs or small family groups. Generally, its food includes rats, mice, lizards, birds' eggs – broken open by being rolled through the hind legs and smashed against rocks – insects, scorpions and wild fruit. The slender mongoose also has a reputation for attacking puffadders, cobras and mambas.

Their main enemies are raptors, so these little creatures keep a constant watch for any ominous shapes in the sky. Should they be attacked, they will often put up a spirited defence. When a ground predator is sighted, suricates (commonly known as meerkats) give a staccato warning bark; this changes to a long, drawn-out sound if a raptor threatens.

Mongooses live in burrows. In the case of the yellow mongoose, the colony can number up to 50 individuals. They are great burrowers, digging tunnels and chambers in sandy soil, or taking over and enlarging a ground squirrel's burrow. The colony can extend over a large area and have up to 100 entrances.

If mongooses are the most ubiquitous creatures, then hyraxes are probably among the most fascinating. Also known as rock rabbit or dassie, the rock hyrax lives in colonies of up to 50 individuals. Each colony is divided into families consisting of a male, several females and young. Within the home range – invariably in some rocky areas – will be special sun-bathing places, drinking spots, look-out points and 'toilet' areas. The entrances to the living quarters are small enough to keep out any interested leopard, jackal, serval or caracal. As an adaptation to their environment, rock hyraxes have pads beneath their feet which, by muscular contraction, can be formed into suction caps. These are kept moist, enabling the animal to climb almost perpendicular cliffs.

Unlike the rock hyrax, the tree hyrax is nocturnal and, like most creatures of the night, occurs in pairs or singly. It lives in holes in trees, often at great heights, but comes down to feed on the ground. Both species are mainly herbivorous, although they sometimes eat insects, lizards and birds' eggs.

Curiously, hyraxes are related to such diverse creatures as elephants and sea cows. In the fossil record, zoologists found that they all had a common ancestor in the Eocene era. After that, evolution made the hyrax smaller, while the elephant got bigger. They are similar in many respects. Small, furry animals usually have paws, but the dassie is different: it has hooves, like an elephant. And both have four toes set quite close together, with blunt plates for nails. The most striking similarity between the two is the set of well developed upper incisors.

The reproductive organs in the two mammals are also alike. Males have their testes inside the body. The females of both species have long gestation periods – 22 months for an elephant, 7½ months for the hyrax – a long time for such a small creature. Like elephants, they also have mammary glands with which they suckle their young.

Two solitary and rarely seen nocturnal creatures are the thick-tailed galago – otherwise known as the bush baby – and his smaller relative, the night ape. Their presence is usually indicated only by their raucous night-time screams. Apart from the difference in size, night apes have bigger ears and eyes than the galago, and are more sociable. In common with both species is a claw-like toe on the hind foot.

Another nocturnal, seldom-seen animal is the ratel or honey badger. This bold, courageous animal has even been known to attack creatures as large as buffalo if annoyed. It has strong jaws and can inflict serious wounds, so it is not surprising that it has few enemies.

Ratels can survive in many types of habitats, from forest to open savanna plains. Their diet is omnivorous, including small animals, rodents, shrews, snakes, insects, spiders, occasionally small antelope, and possibly carrion. They also feed on roots, bulbs, fruits and honey.

In its search for honey, the ratel is helped by the honeyguide. This little bird attracts the ratel to a beehive by its chatter and wing flapping. Should it be necessary, the ratel is capable of climbing a tree to reach the hive. The hive is then smashed and, impervious to bee stings (it is protected by a tough hide and thick fat), the ratel enjoys its meal. As a reward, the honeyguide feeds off the grubs, honey and wax that the ratel discards.

Otters are distant relatives of the ratel. They both belong to the family Mustelidae, which has representatives all over the world in the form of weasels, skunks, stoats, ferrets, minks, polecats and badgers.

There are two species in southern Africa: the spotted-necked otter and the Cape clawless otter. Apart from the differences suggested by their names, clawless otters have only partly webbed feet, so they are not as agile in the water as their relatives. These otters occur over most of southern Africa, except in Namibia's desert regions. Their most favoured habitats are tidal lagoons or pools along the coast. Spotted-necked otters are not found in the southern Cape or Namibia, and are most likely to be seen east and north of Port Elizabeth.

Cape clawless otters have the distinction of being the largest otter species outside South America. Being mainly nocturnal, they are seldom seen, and usually occur in secluded places. Otters hunt crabs, shellfish, frogs, fish – even the big water leguaan. Given the opportunity, they will attack water birds, and even travel considerable distances from water if there is a chance of finding domestic poultry.

The only seal to breed on the coastline of the sub-continent is the Cape fur seal. It inhabits desolate coastal areas and the rocky offshore islands from the Namibian coast to Algoa Bay. On land it is a gregarious animal. Males keep harems of up to 30 females in crowded colonies which can number hundreds of individuals.

As the gestation period is about 11 months, only one pup is born each year. Mating, either on land or in the water, resumes about a week after the birth. Although the young suckle for a year, they can catch squid and small fish by the time they are six months old, by which time they have developed permanent teeth.

These marine carnivores feed on fish such as mackerel, sardines, squid and crabs. To procure their food, they can dive to depths of 50 m, remaining submerged for several minutes. Sharks are their main enemies.

Today, seal breeding colonies are greatly reduced from their large numbers of last century. Yet highly controversial culling programmes are still carried out. The reasons for cropping are similar to those advanced for land mammal culling, except that they are fraught with contradictions, and cannot be scientifically verified. Marine conservation will never have the accurate data available to game reserves simply because it is impossible to assess the numbers of marine species – be they whales, dolphins or seals. Ostensibly, seals are culled because of their competition with Man for fish stocks. They require 2-3 kg of fish daily. Yet their impact on fish resources has been suggested by some researchers to be smaller than that of seabirds.

A complicating factor in seal culling is the revenue derived from selling pelts. Culling is a commercial venture open to exploitation. The issue will probably never be resolved. If it is necessary, then it should be done by a recognised conservation authority, but to determine its necessity there needs to be far more scientific research.

Man is only on the threshold of learning about his world. As yet, we have not recorded all the forms of life on the planet; yet our impact on the biosphere worsens yearly. Unknown species with unknown functions become extinct before science has a chance to record them. Unique habitats are destroyed, be it for agricultural or industrial purposes, before we know the importance of their flora and fauna.

It is both impossible and undesirable to halt mankind's development until the natural scientists have completed their work. The biosphere is resilient, and it is inevitable that we will modify and transform it. But what we *can* ensure is that the way we use the biosphere gives the present generation the greatest sustainable benefit while maintaining its potential to meet the needs and aspirations of future generations.

The suricate, otherwise known as grey meerkat, is often seen squatting on its haunches, surveying its surroundings and keeping a sharp lookout for raptors. Its long claws are ideally adapted for burrowing, though it will often occupy a warren dug by a ground squirrel or yellow mongoose.

A Yellow-billed oxpecker searches for ticks on a warthog. Apart from cleaning their hosts of ticks, oxpeckers warn of approaching danger. Because warthogs are short-sighted, they are quick to respond to the warning calls of birds and other mammals.

Below: A flock of skimmers in flight over the Okavango River. These long-winged, fast-flying birds catch small fish as they skim across the water.

Above: The warthog's social unit is the 'sounder' – usually a family group of adult male and female with her offspring, a maternity group of adult females with sub-adults, juveniles and yearlings, or a bachelor group of adult and younger males. During the night the sounder lies up in an abandoned ant bear hole, which is often enlarged. In addition to providing protection during bad weather, it serves as a refuge from predators. However, lions have been known to dig warthogs out of shallow holes.

Right: Warthogs are enthusiastic mud-wallowers, using the mud coating as a defence against flies and as a means of keeping cool. When water is available, they drink regularly, but otherwise they are largely independent of water.

The ranges of the two species of galago, the thick-tailed bushbaby (right) and the lesser bushbaby (bottom) barely overlap in southern Africa, since the former favours forests and thickets and the latter is a savanna woodland species. Both are nocturnal and most active between sunset and midnight, when they forage for acacia gum, occasionally augmenting their diet with insects. To communicate, galagoes chatter or give vent to a piercing call.

The chicks of a white-backed duck keep among the reeds for protection. Okavango Delta.

Above and left: The Cape clawless otter is predominantly aquatic, occurring in rivers, lakes, swamps, dams, tidal lagoons and pools. It ventures into the sea to hunt crabs and octopuses, but always returns to more sheltered waterways to rest.
Clawless otters are playful animals, often chasing each other around in the water, or indulging in mock fights. Even a solitary otter will play with a stick or stone, sometimes for long periods. A common 'game' is to repeatedly drop a small stone into the water, catching it before it sinks to the bottom.

Middle: The scrub hare is a nocturnal, solitary creature of the savanna woodland. During the day it lies up in a 'form', ideally camouflaged by its drab colouring. If wounded, the scrub hare squeals loudly, kicking and biting if handled. Many captured specimens have been found to lack the last few tail vertebrae, suggesting that intraspecific fighting may be common.

Tree squirrels are active during the early and late part of the day, and may be seen in pairs, small family parties or (occasionally) alone. Although predominantly vegetarian, they supplement their diet with insects – especially during the dry seasons, when berries, leaves and seeds may be scarce.Transvaal lowveld.

When feeding on grasses, seeds, bulbs or roots, the ground squirrel carries its tail like an umbrella to shade itself from the sun.

Right: Ground squirrels are born blind and naked in a warren made up of a complicated system of burrows. They are covered in hair within 14 days, but do not open their eyes until after 35 days. After about 40 days, they make short excursions from the burrow, and after 50 days they start weaning. A group of ground squirrels is dominated by a female which vigorously defends the territory around the tunnel opening.

Ground squirrels (far left) suricates (left) and a yellow mongoose (above) frequently share a warren, although the first two species will ignore one another and keep to distinct parts of the burrow system. However, yellow mongooses and ground squirrels may live in harmony for long periods, and can sometimes be seen sunning themselves together. Nevertheless, a yellow mongoose will kill and eat a sick or injured ground squirrel.

Top left, top right & bottom left: Dwarf mongooses are the smallest of the mongoose species found in southern Africa. These diurnal creatures form packs of 8-10 individuals which usually take up permanent residence in abandoned termite mounds (as shown in top right and bottom left). However, they will also excavate their own burrows. When foraging in the morning they scatter over a wide area, maintaining contact by issuing a series of high-pitched sounds.

Top middle: Selous' mongoose (shown here in the Makgadikgadi Pans Game Reserve) is a savanna species which favours scrub and woodland. During the day it lies up in a self-excavated burrow, which may have more than one entrance, venturing out at night to prey on rats, mice, small birds and reptiles.

Above: Banded mongooses are gregarious, living in troops that may number only four or five or more than 30 individuals. Their acute sense of smell enables them to locate grubs underground, and they are quick to detect the movements of grasshoppers and other insects by sight. Although they are mainly insectivorous, these mongooses will also eat lizards, snakes and wild fruit.

Right: The solitary slender mongoose occurs in habitats ranging from arid areas to regions where the annual rainfall is higher than 1400 millimetres. Although normally found on the ground, it is a good climber and may take to trees when hunting.

Far right: Ground squirrels occur throughout the more arid parts of southern Africa. They form groups of several females with their offspring, the males moving constantly between the groups – remaining only if a female is in oestrus.

Top, above and right: Two types of hyrax are found in southern Africa, the rock hyrax and the tree hyrax. However, their names are no help in identification, as both types can be seen on rocks and in trees. Sunning themselves on a rock are two tree hyraxes (top left), while a yellow-spotted rock hyrax (right) crouches in a tree. The reddish-brown pair (above) are also rock hyraxes.

Of the eight subspecies of tree hyrax, only one occurs in the sub-continent, and then only in forested parts of the eastern Cape, Natal and Zimbabwe. It is mostly nocturnal (whereas the rock hyrax is predominantly diurnal). This hyrax has an extraordinary call which begins as a soft groan and rises to an ear-splitting shriek.

Unlike most mammals, rock hyraxes are only poorly able to regulate their body temperatures. This is why they can often be seen basking in the sun to absorb heat. At night, when the temperature falls, they huddle together in rocky shelters to retain warmth.

The yellow-spotted rock hyrax is a subspecies which occurs throughout eastern Africa, but no further south than the northern Transvaal. There is a high degree of individual colour variation, and the dorsal spot may be creamy or yellow.

Top left: Cape fur seal pups explore the rocks and muddy pools of Baker's Bucht along the southern Namib desert coast.

Above & Right: The Cape fur seal is the only seal that breeds on the southern African coastline, where it occurs along the desolate Namibian seaboard and around the Cape to Algoa Bay. On land it lives in crowded, noisy colonies which can number thousands of densely packed individuals.

Far Right: Spilling down to the beach, a large group of seals makes for the turbulent Atlantic breakers. Seals appear ponderous and slow-moving on land, but in water they are agile swimmers and divers, capable of remaining submerged for up to 20 minutes.

PORTRAITS OF THE WILD

Index